Wettability and Interfacial Phenomena - Implications for Material Processing

Edited by Rita Khanna

Published in London, United Kingdom

IntechOpen

Supporting open minds since 2005

Wettability and Interfacial Phenomena - Implications for Material Processing
http://dx.doi.org/10.5772/intechopen.75334
Edited by Rita Khanna

Contributors
Qiaoli Lin, Ran Sui, Naofumi Hiraoka, Ivaldo Ferreira, Amauri Garcia, J.A. Castro Castro, Ghada Bassioni, Syed Taha Taqvi, Lacramioara Popa, Cristina Elena Dinu-Pîrvu, Mihaela Violeta Ghica, Roxana-Elena Avrămescu, Xiaogang Yang, Lu Liu, Xiani Huang, Chenyang Xue, Guang Li, Yeeli Kelvii Kwok, Rita Khanna

First published in London, United Kingdom, 2019 by IntechOpen
IntechOpen is the global imprint of INTECHOPEN LIMITED, registered in England and Wales, registration number: 11086078, The Shard, 25th floor, 32 London Bridge Street
London, SE19SG – United Kingdom
Printed in Croatia

British Library Cataloguing-in-Publication Data
A catalogue record for this book is available from the British Library

Additional hard copies can be obtained from orders@intechopen.com

Wettability and Interfacial Phenomena - Implications for Material Processing
Edited by Rita Khanna
p. cm.
Print ISBN 978-1-83880-538-8
Online ISBN 978-1-83880-539-5
eBook (PDF) ISBN 978-1-83880-713-9

For EU product safety concerns:
IN TECH d.o.o., Prolaz Marije Krucifikse Kozulić 3, 51000 Rijeka, Croatia, info@intechopen.com or visit our website at intechopen.com.

Meet the editor

Associate Professor Rita Khanna holds a PhD in Condensed Matter Physics and has extensive research experience in the fields of diffuse X-ray scattering, disordered materials, atomistic computer modeling and simulations, high-temperature metallurgical processes, environmentally sustainable waste management, nanoscale carbons, foams, structures along with extensive wettability investigations on carbon/molten iron, carbon/slag interactions, carbon-based oxide refractories, and computer modeling at high temperatures. She has worked at several leading institutes: the Nuclear Research Centre, Jülich, Germany; the Department of Atomic Energy, India; the Australian National University, Canberra; Sydney University; and the University of New South Wales, Sydney, Australia. She has over 153 publications to her credit: 1 edited book, 12 book chapters, 112 journal and 28 conference papers.

Contents

Preface

Dynamical changes and modulations in surface wetting and surface tension-related processes across solid/liquid interfaces can have a significant impact on overall reaction kinetics and have been the focus of extensive research in physical, chemical, biological, agricultural, and environmental sciences. Some of the key applications include studies on groundwater flows, oil spills, water management, disease transmission, chemical leaching, nanotechnology, etc. This book presents a wide spectrum of studies in the fields of nanomaterials, oil recovery in the petroleum industry, metal processing, surface tension and viscosity of fluids, and superhydrophobic behaviour in a variety of natural phenomena.

The book has an introductory chapter and seven full chapters with contributing authors from Australia, the People's Republic of China, Japan, Brazil, Egypt, Canada, Hong Kong, and Romania. Global authorship in this book reflects multifaceted interest and activity in this field worldwide with breakthroughs occurring in several research areas.

The table of contents is organized considering novel applications of wettability and interfacial phenomena to different fields, including practical treatments to enhance system behaviour. These include nanoscale wettability, interactions of oil with metals and reservoir rocks for applications in the petroleum industry, the role of coatings, particle morphology, surface roughness and viscosity in metal processing, and industrial applications of superhydrophobic behaviour.

The introductory chapter presents an overview of factors influencing the wettability of nanomaterials, a key and emerging area of wettability research. Nanoscale surface interactions between graphene, carbon nanotubes, and other materials with water, aqueous solutions, and polymers can play a significant role in water filtration, desalination, and polymer dispersion applications. Key factors that could be used to modify and control nanoscale wetting behaviour are discussed along with novel wetting investigations on nanochannels, membranes, and nanostructures.

The second chapter reports on the wetting behaviour of oil-impregnated sintered bearings and friction reduction for applications in automobile parts, office automation equipment, motor vehicles, etc. Shaft surface oil wettability was investigated for PTFE-coated shaft-impregnated sintered bearings and compared to noncoated shafts as an effective means of reducing friction. With oil from the porous bearing body lubricating the shaft and bearing surfaces, the low oil-wettable shaft was found to retain a larger amount of oil in the bearing clearances and indicated lower friction than a highly wettable shaft.

The third chapter reports on determining a complete wettability profile based on zeta potential measurements on reservoir rocks, one of the primary factors that control location, saturation, distribution, and flow behaviour of reservoir fluids.

Factors such as temperature, pressure, and pH can have an effect on reservoir rock wettability, which in turn influences several petrophysical properties such as capillary pressure, relative permeability, waterflood behaviour, electrical properties, and enhanced oil recovery. Detailed wettability studies are reported on limestone rock in contact with crude oil and asphaltenic solutions of varying concentrations in limestone/water suspension.

The fourth chapter reports wetting investigations on the coating of steel and titanium alloys with two aluminium alloys aimed at enhancing anti-corrosion, service time, and working temperatures for hot dipping, coating, casting, and brazing or soldering processes. The wettability of Al/steel and Al/Ti systems was found to improve with the formation of intermetallics at the liquid/solid interface, and the reduction of precursor oxide film. Thermodynamics of solute segregation was also discussed toward developing relevant wetting theories.

The fifth chapter reports on the influence of alumina surface processing on the surface morphology and wetting behaviour of molded steel as determined with 3D profilometer and scanning electron microscopy. Processing time was found to play a critical role in surface roughness decreasing with increasing time but with the wettability of the processed surface changing from hydrophilic to hydrophobic. Wettability was found to reverse to hydrophilic with further increases in processing times.

The sixth chapter presents a report on the effect of fluid dynamics based on the morphology and sizes of barium sulfate particles in a Taylor-Couette flow reactor. Various changes in particle morphology were found to depend on the hydrodynamics, the onset of turbulent flow, supersaturation, feeding rates, and the rotational speed of the inner cylinder of the reactor. While micro-mixing due to fluid shear had a strong influence on size dispersion and morphology, supersaturation played a key role in crystallization and particle agglomeration.

The seventh chapter reports a theoretical investigation on calculating the viscosity of aluminium alloys based on surface tension/viscosity relation equations for pure metals and derives a unified solution for viscosity of multicomponent alloys. The derived solution for surface tension was computed as a function of temperature for ternary Al–Cu–Si and Al–Cu–Si–Mg alloys. With increasing alloy Si content, the surface tension of Al-based alloys was found to decrease in good agreement with the casting practice of aluminium-based alloys.

The eighth chapter presents an interdisciplinary approach on extreme wettability with specific attention to superhydrophobic natural and artificial surfaces and to liquid marbles. Studies reveal a "non-wetting" contact with solid supports and many unexpected properties, such as versatility in choice of cores and shells, recoverable deformability, ability to float on water, low evaporation rate, etc. The super-hydrophobicity concept is finding application in anti-bacterial or self-cleaning fabrics, cell culturing/screening/isolation, and anti-icing, anti-reflective, and anti-corrosion materials/coatings, which are gaining attention in the transportation and optical devices fields.

This book covers extensive areas of interest in wetting and wettability, interfacial reactions, applications to several practical problems, and recent developments in the field. The book has a global perspective and wide coverage of topics for academics, professionals, and regional and international organizations.

Rita Khanna
Associate Professor,
School of Materials Science and Engineering (Retd.),
The University of New South Wales,
Sydney, Australia

Chapter 1

Introductory Chapter: Factors Influencing the Wettability of Nanomaterials

Rita Khanna

1. Introduction

Surface wetting, capillarity, adhesion, and surface tension-related processes across solid-liquid interfaces have been the focus of extensive theoretical and experimental research in fields such as natural sciences, agriculture, geophysics, technology, water management, biological, and environmental sciences. Intermolecular interactions play a key role in the ability of a liquid to maintain contact with a solid surface, and a balance between cohesive and adhesive forces determines the overall wetting behavior of the system. Some of the wettability applications include superhydrophobic surfaces, dynamics of oil spills, ground water flows, disease transmission, chemical leaching, nanotechnology, and several other real-life applications [1, 2].

The wetting behavior of nanomaterials such as carbon nanotubes (CNT), graphene, graphyne, nanoparticles, and nanoengineered surfaces is an area of intense experimental and theoretical research activity [3–5]. Intermolecular interactions are of crucial importance for controlling nanoscale material behavior in various aspects of nanotechnology, nanodevices, and their applications. Other areas of research include the flow of liquids inside nanochannels, tuning of nanotube forests and arrays for modifying wetting characteristics, development of nanogrippers for manipulating carbon nanotubes for electro-mechanical devices, nanoscale surface treatments for producing hydrophilic or superhydrophobic surfaces [6].

Several advanced optical or microscopic experimental techniques are being used for nanoscale wetting investigations. While the macroscopic contact angle at the solid-liquid interface can be measured using conventional optical techniques, advanced microscopic or indirect techniques are required for micro and nanoscale investigations. It is also a common practice to determine an "effective contact angle" at nanoscale distances due to limited/poor contact at the liquid meniscus, confinement issues, and the influence of long-range forces [7].

A brief overview of measurement techniques is provided next. Surface forces technique determines the influence of separation distance on forces (±10 nN) between two surfaces with the help of capacitive sensors or springs; distances can be controlled down to 0.1 nm with help of piezoelectric positioners. Wilhelmy method measures the force exerted during contact of liquid with a solid specimen for an indirect determination of contact angle [8]. This technique has been used to determine the wetting behavior of nanowires (500 nm dia.) and nanoneedle probes. Widely used sessile drop technique determines contact angle directly from the profile of a liquid drop atop a solid substrate with help of video cameras or telescopic arrangements. This technique can be used to continuously monitor dynamic changes in contact angles as a function of various system variables [9].

Interference microscopy technique computes contact angles (15–30°) from the 3D contours and profiles of liquid droplets by using interference fringe patterns formed at the solid/liquid and liquid/vapor interfacial region [10]. Atomic force microscopy scans the sample surface with a sharp tip to monitor surface topology while maintaining a constant interaction force through feedback loops; a modified nanodispenser tip is used for micro- and nanodroplets [11]. Scanning Polarization Force Microscopy uses electric polarization forces for mapping topography contours for distances in the range 10–20 nm, thereby minimizing the risk of surface contact by probing tip, an aspect especially valuable for soft and liquid specimens [12].

Extensive theoretical research is also being carried out using theoretical modeling, analysis, Monte Carlo, and molecular dynamics computer simulation on a variety of relevant issues. Some of the problems being investigated include the influence of nanoscale confinement on surface tensions, wetting behavior inside small capillaries and nanochannels, effect of line tension, liquid adsorption on capillary walls, and precursor films, the role of nanoscale surface roughness on confinement and wetting among others [13, 14].

In this chapter, a brief overview on the nanoscale wetting behavior of graphene and carbon nanotubes is presented along with factors influencing the wetting phenomena and basic system characteristics of nanomaterials.

2. Graphene

Remarkable technical developments, based on carbon's ability to bind to itself in a variety of hybridized states, have seen the emergence of synthetic carbons with exotic and exceptional properties, e.g., fullerenes, nanotubes, graphene, graphyne, nanorings, etc. Graphene, the ultimate thin material, is composed of a single sheet of graphite, and is currently one of the most investigated 2D carbon allotrope [15]. Its existence was considered to be purely theoretical for a long time. Single graphene layers were prepared in 2004 by a simple mechanical exfoliation of graphite using a Scotch tape. Graphene is finding extensive applications in nanotechnology, optoelectronics, water desalination, dispersion in polymers, etc. Several of these applications, such as capacitive energy storage and heat transfer coefficient, depend strongly on graphene's wettability and its surface-based interactions with various liquids [16].

With all its carbon atoms exposed to high levels of surface activity, graphene is a material of choice for interfacial carbon materials, especially for carbon-aqueous liquid interactions. Poor wettability of graphene with water and its hydrophobic nature can limit device contamination during fabrication; however, there are concerns regarding the hydrophilic or hydrophobic nature of graphene and its dependence on operating parameters. The contact angle of graphene with water and aqueous solutions has been found to vary depending upon various conditions, e.g., surface wrinkle morphology, extent of tensile strain, vibrational strain, etc. Other variables include the influence of doping, presence of defects, temperature, and electric field, among others. Graphene monolayer also finds application as an ideal coating material due its wetting transparency [17].

Vertical graphene sheets for applications, such as high-performance super capacitors, are presently being produced using plasma-enhanced chemical vapor deposition technique. During sputtering and deposition process, a number of defects get invariably incorporated in graphene sheets. The density of defects can be controlled through sputtering time. Graphene with low density of defects showed a hydrophobic behavior and poor wettability; significant improvements in the wetting were, however, observed for high defect concentrations [18]. Laser-induced graphene is produced by irradiating polyimide film with CO_2 laser to burn away all elements except carbon. Operating conditions during their production can impact the chemical composition of the graphene surface affecting its wettability with water; the presence of H_2 gas in the chamber along with limited O and CO contents led to hydrophobic graphene [19].

Graphene, typically a zero-band-gap semiconductor, can be doped as n-type or p-type using sub-surface polyelectrolyte or metals, electrical voltage chemicals, etc., and its wetting properties can be modified with significant changes in contact angles [20]. Doping can modify and regulate the surface electron density of graphene and its interaction with external molecules in the contact region. It may also help to develop a feasible route for tuning the surface wettability of graphene, especially for coating applications. Studies have also been reported on the conversion of its surface wettability from hydrophobic to hydrophilic and vice versa with help of external stimuli [21]. Extensive research is being carried out in this field on several aspects of the wetting behavior of graphene with focus on controlled and reversible tuning.

3. Carbon nanotubes (CNTs)

Carbon nanotubes, seamless rolled-up graphite sheets, are tubular networks of sp^2-hybridized carbon atoms; these 2D structures are considered to be one-dimensional due to their high aspect ratio (1–2 nm dia., length in mm) [22]. Produced using techniques such as chemical vapor deposition, graphite evaporation, etc., these can either be single- or multi-walled depending on the production route [23]. Due to several unusual characteristics including nanoscale diameters, non-polar surfaces, CNTs are being investigated for nanoscale water desalination and selective transport in nanochannels. The wetting behavior of CNTs, nanotube forests and arrays, effects of nanoscale confinement, and tunability are areas of intense research [24, 25].

The effect of various conditions on the wetting behavior of CNTs with water has been investigated by several researchers. Key factors were found to be the structure of CNTs, tube's size, the strength of van der Waals interaction between water and carbon, spontaneous imbibition of water in CNTs, surface energies, and temperature [26, 27]. MD simulations on SWCNTs in the temperature range 270–370 K showed an increase in capillary filling and water uptake with increasing temperature. This was attributed to increased wettability and reduced viscosity of water with temperature along with small increases in water imbibition [28].

The wetting behavior inside small capillaries or nanochannels can be very complex depending on intermolecular interactions between the solid and the liquid, van der Waals attraction, surface tension, line tensions along the radius of the nanotubes, tube diameter, etc. [29]. The adsorption of liquid atoms as mono/multilayers on capillary/CNT walls can distort the meniscus profile during the filling of nanochannels and may affect the wetting behavior due to disjoining pressure induced by molecular interactions. Surface roughness can also play an important role during nanoscale confinement, especially when the characteristic variations in surface roughness become comparable to the channel diameters. Surface chemistry is another key factor affecting wettability and flow through nanochannels [30].

A number of investigations have been reported on "nanopumping" to pump liquids through nanotubes based on electrical or mechanical properties of CNTs including studies on the flow of viscous fluids, activation of fluid flow, friction with nanowalls, etc. [31]. The wetting behavior of CNTs has been investigated with a number of polymers, e.g., polyvinylidene fluoride, maleic acid anhydride, polycaprolactone, polyurethane acrylate, etc. Chemical surface treatments were used to modify the wettability of nanochannels [32]. The wetting of nanotube forests and arrays has been investigated with several liquids including water, glycerol, ethylene glycol, propylene carbonate, olive oil, dimethyl sulfoxide, and nitromethane with contact angles ranging between 60 and 157° [33]. CNT forests have been used to grow hematite nanochain arrays, as filters to capture micro/nanocontaminants in water, and as gas diffusion arrays with contact angles up to 150° [34].

4. Concluding remarks

A number of factors may be used to affect, control, and tune the wettability of nanomaterials. There is immense potential for creating materials with designer characteristics tailored to specific applications. The influence of wetting characteristics, including contact angles, chemical composition of nanosurfaces, and surface roughness on the flow behavior of water, aqueous solutions, and polymers, is being investigated for water filtration, desalination, dispersion in polymers, etc. The contact angle of graphene with water has been found to depend upon doping, concentration of defects, temperature, electric field, surface wrinkle morphology, tensile, and vibrational strains. Nanoscale wetting investigations on materials, e.g., CNTs, graphene, nanochannels, membranes, and nanostructures represent an area of intense theoretical as well as experimental research activity.

Author details

Rita Khanna
School of Materials Science and Engineering (Retd.), UNSW Australia, Sydney, NSW, Australia

*Address all correspondence to: rita.khanna66@gmail.com

References

[1] Yuan Y, Randall LT. Contact angle and wetting properties. In: Bracco G, Holst B, editors. Surface Science Techniques. Springer Series in Surface Science. Berlin/Heidelberg: Springer-Verlag; 2013. 3p

[2] Beebe DJ, Mensing GA, Walker GM. Physics and applications of microfluidics in biology. Annual Review of Biomedical Engineering. 2002;**4**:261-286

[3] Bahrami AH, Jalali MA. Nanoscopic spontaneous motion of liquid trains: Nonequilibrium molecular dynamics simulation. Journal of Chemical Physics. 2010;**132**:024702

[4] Jing FJ, Guo Z. Wettability of graphene: From influencing factors and reversible conversions to potential applications. Nanoscale Horizons. 2019;**4**:339-364

[5] Khanna R, Ikram-Ul-Haq M, Rawal A, Rajarao R, Sahajwalla V, Cayumil R, et al. Formation of carbyne-like materials during low temperature pyrolysis of lignocellulosic biomass: A natural resource of linear sp carbons. Scientific Reports. 2017;**7**:16832

[6] Costanzo PM, Wu W, Giese RF, van Oss CJ. Comparison between direct contact angle measurements and thin layer wicking on synthetic monosized cuboid hematite particles. Langmuir. 1995;**11**:1827-1830

[7] Moldovan A, Enachescu M. Wetting properties at nanometer scale. In: Aliofkhazraei M, editor. Wetting and Wettability. Croatia: IntechOpen; 2015. 15p

[8] Eral HB, Mannetje DJCM, Oh JM. Contact angle hysteresis: A review of fundamentals and applications. Colloid and Polymer Science. 2013;**291**:247-260

[9] Srinivasan S, McKinley GH, Cohen RE. Assessing the accuracy of contact angle measurements for sessile drops on liquid-repellent surfaces. Langmuir. 2011;**27**:13582-13589

[10] Sundberg M, Månsson A, Tågerud S. Contact angle measurements by confocal microscopy for non-destructive microscale surface characterization. Journal of Colloid and Interface Science. 2007;**313**:454-460

[11] Salmeron M. Nanoscale wetting and de-wetting of lubricants with scanning polarization force microscopy. In: Bhushan B, editor. Fundamentals of Tribology and Bridging the Gap Between the Macro- and Micro/Nanoscales. NATO Science Series II. Vol. 10. Dordrecht: Kluwer Academic Publishers; 2001. pp. 651-662

[12] Luna M, Rieutord F, Melman NA, Dai Q, Salmeron M. Adsorption of water on alkali halide surfaces studied by scanning polarization force microscopy. Journal of Physical Chemistry A. 1998;**102**:6793-6800

[13] Chen J, Hanson BJ, Pasquinelli MA. Molecular dynamics simulations for predicting surface wetting. AIMS Materials Science. 2018;**1**:121-131

[14] Wan JM, Wilson JL. Colloid transport in unsaturated porous media. Water Resources Research. 1994;**30**:857-864

[15] Segal M. Selling graphene by the ton. Nature Nanotechnology. 2009;**4**:612-614

[16] Taherian F, Marcon V, van der Vegt NFA, Leroy F. What is the correct contact angle of water on graphene? Langmuir. 2013;**29**:1457-1465

[17] Preston DJ, Mafra DL, Miljkovic N, Kong J, Wang EN. Scalable graphene

coatings for enhanced condensation heat transfer. Nano Letters. 2015;**15**:2902-2909

[18] Stoller MD, Park S, Zhu Y, An J, Ruoff RS. Graphene-based ultracapacitors. Nano Letters. 2008;**8**:3498-3502

[19] Li Y, Luong DX, Zhang J, Tarkunde YR, Kittrell C, Sargunaraj F, et al. Laser-induced graphene in controlled atmospheres: From superhydrophilic to superhydrophobic surfaces. Advanced Materials. 2017;**29**:1700496

[20] Hong G, Han Y, Schutzius TM, Wang Y, Pan Y, Hu M, et al. On the mechanism of hydrophilicity of graphene. Nano Letters. 2016;**16**:4447-4453

[21] Yao X, Song Y, Jiang L. Applications of bio-inspired special wettable surfaces. Advanced Materials. 2011;**23**:719-734

[22] Iijima S, Ichihashi T. Single-shell nanotubes of 1-nm diameter. Nature. 1993;**636**:603-605

[23] Lau PH, Takei K, Wang C, Ju Y, Kim J, Yu Z, et al. Fully printed, high performance carbon nanotube thin-film transistors on flexible substrates. Nano Letters. 2013;**13**:3864-3869

[24] Starov VM, Velarde MG, Radke CJ. Wetting and Spreading Dynamics. Surfactant Science Series. Boca Raton, Florida: CRC Press; 2007

[25] Dujardin E, Ebbesen TW, Hiura H, Tanigaki K. Capillarity and wetting of carbon nanotubes. Science. 1994;**265**:1850-1852

[26] Holt JK, Park HG, Wang Y, Stadermann M, Artyukhin AB, Grigoropoulos CP, et al. Fast mass transport through sub-2-nanometer carbon nanotubes. Science. 2006;**312**:1034-1037

[27] Vaitheeswaran S, Rasaiah JC, Hummer G. Electric field and temperature effects on water in the narrow nonpolar pores of carbon nanotubes. The Journal of Chemical Physics. 2004;**121**:7955-7965

[28] Ebrahimi F, Moghaddam MG. Temperature-dependence of wetting properties of carbon nanotubes. Physica A. 2016;**453**:271-277

[29] Rossi MP, Gogotsi Y, Kornev KG. Deformation of carbon nanotubes by exposure to water vapor. Langmuir. 2009;**25**:2804-2810

[30] Joseph S, Aluru NR. Why are carbon nanotubes fast transporters of water? Nano Letters. 2008;**8**:452-458

[31] Gong X, Li J, Lu H, Wan R, Li J, Hu J, et al. A charge-driven molecular water pump. Nature Nanotechnology. 2007;**2**:709-712

[32] Liu M, Jia Z, Liu F, Jia D, Guo B. Tailoring the wettability of polypropylene surfaces with halloysite nanotubes. Journal of Colloid and Interface Science. 2010;**350**:186-193

[33] Cho WK, Choi IS. Fabrication of hairy polymeric films inspired by geckos: Wetting and high adhesion properties. Advanced Functional Materials. 2008;**18**:1089-1096

[34] Stuckey P, Lin J, Kannan A, Ghasemi-Nejhad M. Gas diffusion layers for proton exchange membrane fuel cells using in situ modified carbon papers with multi walled carbon nanotubes nanoforest. Fuel Cells. 2010;**10**:369-374

Chapter 2

On the Friction of Oil-Impregnated Sintered Bearings

Naofumi Hiraoka

Abstract

Oil-impregnated sintered bearings are widely used in various products. Friction reduction in them is still a large target for the related industries. In those bearings, lubricating oil exudes from the porous bearing body and lubricates the shaft and bearing surfaces. However, the amount of oil in those sliding areas is often insufficient leading to an unsatisfactory friction. Oil wettability of the shaft surfaces was found to have a large effect on the friction of those bearings. Low oil-wettable shaft could retain a larger amount of oil in the bearing clearances and indicated lower friction than highly wettable shaft. This is because a large contact-angle hysteresis between the oil and the low-wettable shaft surface allows the retention of large oil droplets in the bearing clearances. The control of oil-wettability of the shaft surface could be an effective means of reducing friction for oil-impregnated sintered bearings.

Keywords: friction, lubrication, wettability, surface tension, contact angle, surface treatment

1. Introduction

Oil-impregnated sintered bearings are widely used in various products, especially in motor vehicles and office automation (OA) equipment [1, 2]. For example, they are used in more than 30 parts of automobiles [3]. In these bearings, lubricating oil exuding from the porous bearing body to the sliding surface lubricates the shaft and the bearing. The exudation occurs due to thermal expansion of the oil volume by frictional heat, the "pumping effect" caused by negative pressure in the lubricating oil film [2, 4], and capillary force. Due to such mechanisms, these bearings are expected to have a long service life and relatively low friction without an external oil supply.

Generally, friction of oil-impregnated sintered bearings is larger than fully lubricated bearings, because they are often under boundary lubrication condition. These bearings are often used in relatively small parts, and their friction loss is a serious issue because of small power consumptions allowed for such parts. Frictional noise is also a problem, especially for OA equipment, which is related to lubrication conditions. To improve the lubrication conditions and reduce the friction of these bearings has been the challenge for many researchers and is still a major problem today.

In this chapter, lubrication conditions of oil-impregnated sintered bearings are roughly explained, and the effects of oil wettability of the shaft surface on improving the lubrication conditions are discussed [5].

2. Lubrication conditions

Figure 1 shows the optical microscopic image of the oil-impregnated sintered bearing surface. Many pores and gaps among the grains could be found in sintered metal body. Lubricating oil is impregnated in these pores and gaps usually using vacuum condition. As described above, lubricating oil exudes from the porous body to the sliding surface in some ways during sliding and lubricates the shaft and bearing surface. The schematic view of the shaft and bearing is shown in **Figure 2**. As can be considered from these oil supply mechanisms, oil amount in the bearing gap is usually smaller than fully lubricated bearings and sometimes insufficient for good lubrication.

In addition to oil supply problem, there are some features that make low friction difficult to realize. Some exuded oil returns into the pores of the bearing body by capillary force and also leading to insufficient oil on the sliding surface [4, 6]. The porous surface of the bearing means a reduction in the load area, which lowers the loading capacity of the bearing. Although fluid lubrication was reported to be possible for oil-impregnated sintered bearings especially under large sliding speed conditions [7], the sliding area of the bearing is often under a boundary or mixed

Figure 1.
Optical microscopic image of oil-impregnated sintered bearing surface. Arrows indicate the example of a grain and a pore.

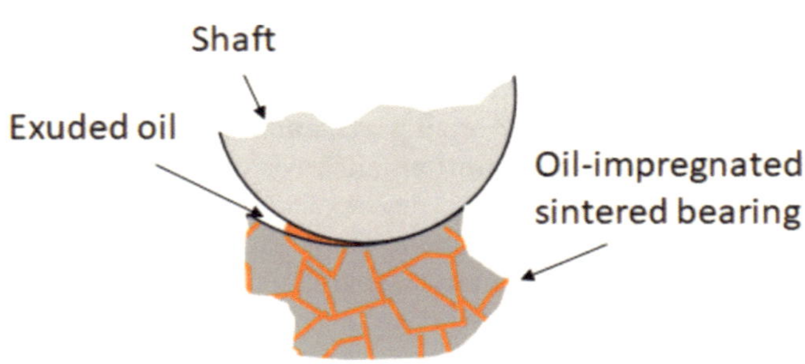

Figure 2.
Schematic view of shaft and bearing.

lubrication condition [6] due to the reasons mentioned above, resulting in large bearing friction.

To improve the bearing friction under such lubrication conditions, solid lubricants are sometimes added to the bearing materials, or an oiliness agent (e.g., stearic acid) is added to the impregnated oil [2, 4, 8]. However, the friction of oil-impregnated sintered bearings is generally higher than that of fully lubricated bearings [6].

3. Oil wettability of shaft surfaces

Metal surfaces generally show good wettability with oils. Lubricating oil could spread easily into sliding area and stay there stably due to this property. This realizes good lubrication condition for machines.

There is a report that a low oil-wettable surface showed low friction due to the slip between the oil and the surface especially under fluid lubrication condition [9]. Some processing, coating, doping, etc., on the surface is usually necessary to provide low oil wettability to metal surfaces and needs additional cost. Recently, a novel grinding method was reported to give processed surface hydrophilicity [10], which probably could give low oil wettability to the surface at low cost. Though such techniques will possibly generalize low oil-wettable shafts, it seems that providing low oil wettability to the metal surfaces is not widely practiced so far to reduce the friction considering the benefits of good wettability to the lubricity described above and cost-effectiveness.

However, as will be shown below, imparting the low oil wettability to the shaft can be a relatively simple way to basically improve the lubrication condition for oil-impregnated sintered bearings.

4. Friction of oil-impregnated sintered bearings

4.1 Friction measurement for shafts with different wettability

Friction coefficients of oil-impregnated sintered bearings with stainless steel shafts were measured. Specimens used are described in **Table 1**. The oil-impregnated sintered bearing used is commercially available mainly for OA equipment, and the size was relatively small.

Bearing	Fe-Cu-based commercial oil-impregnated sintered bearing Oil content: 20.4~20.8% Bore: 4.009 mm. Width: 3 mm
Impregnated oil	Synthetic oil Kinematic viscosity: 42.9 mm^2/s@40°C Surface tension: 25 mN/m@20°C
Noncoated shaft	Material: JIS SUS420J2 hardened stainless steel Diameter: 3.996~3.997 mm Surface roughness: Ra 0.022~0.028 μm
PTFE-coated shaft	Material: JIS SUS420J2 hardened stainless steel + PTFE transfer coating Diameter: 3.996~3.997 mm Surface roughness: Ra 0.024~0.028 μm

Table 1.
Specimen descriptions.

A simple test rig shown in **Figure 3** was used to measure the frictional torque of the bearings. The bearing was inserted and fixed in a ring-shaped weight. The frictional torque T was measured and converted to the friction coefficient μ by

$$\mu = T/(RW) \tag{1}$$

where R is a radius of the bearing bore, and W is the weight load.

Tests were conducted under a weight load of 6.7 N and at a shaft rotation speed of 100–800 rpm. Average frictional torque values were measured every 10 s for each 100-rpm increase in speed. Note that bearing frictional torque is usually different from that of the shaft.

Two types of the shafts: noncoated and PTFE coated stainless steel shafts in **Table 1**, were evaluated. The noncoated stainless steel shafts were used for oil wettability shaft. For the low wettability shaft, PTFE (polytetrafluoroethylene) transfer film coated stainless steel shafts were used. The PTFE block was lightly pressed against the rotating shaft surface and then gently wiped with a nonwoven fabric to make a PTFE transfer coating. As PTFE is known for its low surface energy and ability to make a transfer film on the counter surface by sliding [11], the shaft surface made by this method was expected to have low oil wettability.

Surface roughness was not much different between the stainless steel shaft and the PTFE coated shaft surface as shown in **Table 1**. The shaft diameter measurement by a micrometer before and after application of the coating indicated the PTFE coating thickness was less than 1 μm, though exact thickness was unknown. The past literatures have showed the PTFE transfer film thickness was in the order of several nm [12–14].

Oil droplets on the shafts about 10 s after being deposited there were shown in **Figure 4**. Colors of the droplets appeared a little different between the shafts, but it was due to the oil droplet thickness. Contact angle for noncoated shaft was much lower than that of the PTFE-coated shaft. The oil droplet on the noncoated shaft subsequently spread to cover the wide range of the shaft surface, while that on the PTFE-coated shaft retained its original droplet shape. This clearly indicates the lower oil wettability of the PTFE-coated shaft surface, as compared to that of the noncoated shaft surface.

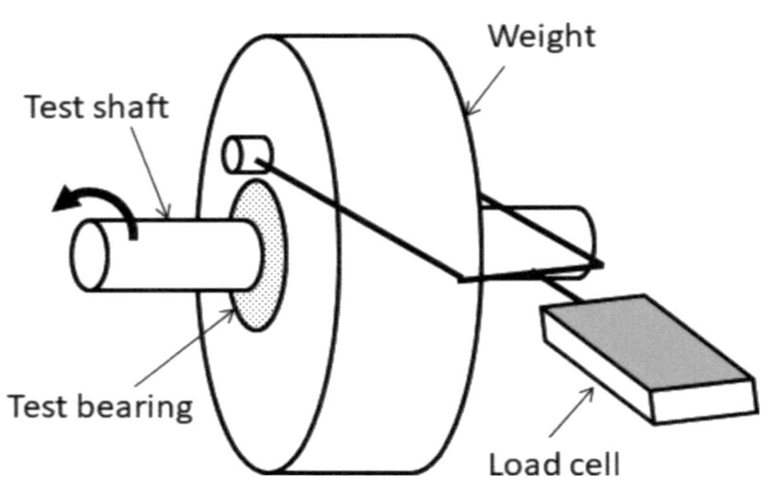

Figure 3.
Schematic view of bearing friction test rig.

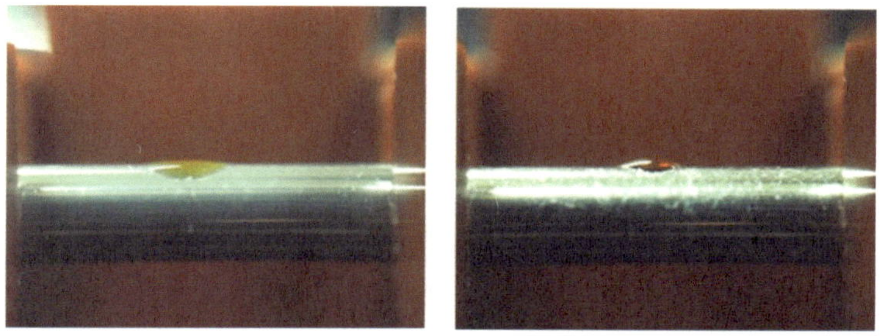

Figure 4.
Oil droplets on the shafts.

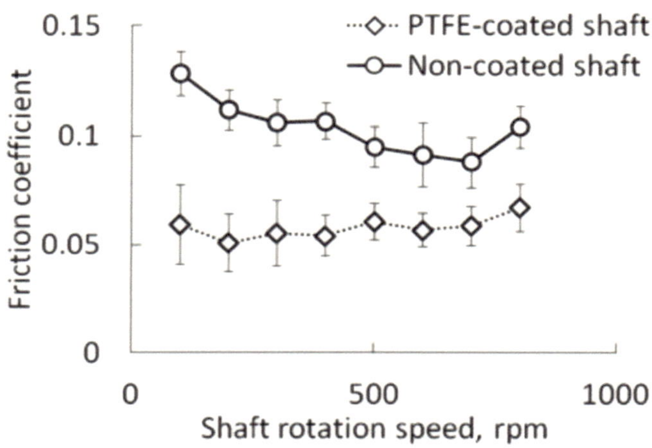

Figure 5.
Friction coefficients of the bearings. Error bars indicate the standard error of the mean.

Figure 5 shows the friction coefficients of the bearings calculated by Eq. (1) with PTFE-coated and noncoated shafts. Error bars indicate the standard error of the means of five time test data for each shaft. The PTFE-coated shaft indicated a friction coefficient 20–50% lower than that of the noncoated shaft on average. Lubrication condition was presumed to be boundary lubrication, because friction coefficients are not largely dependent on shaft rotation speed.

Figure 6 shows metallurgical microscopic (Olympus BHMJ, Japan) images of typical shaft and bearing surfaces before and after the tests. As compared to the intact shaft surface which could be estimated from the nonsliding area, wear marks were found on the sliding areas of shafts and bearings of both non- and PTFE-coated shaft tests. Comparing these surfaces, those for the PTFE-coated shaft test showed comparatively milder wear appearance.

4.2 Oil amount in the bearing clearance and the shaft wettability

As was described, one of the big problems for the lubricity of oil-impregnated sintered bearings is poor oil supply to the sliding surfaces. We focused on the oil amount on the shaft and in the bearing clearance first as a cause of the friction reduction of the PTFE-coated shaft.

Figure 7 shows oil deposition on the shafts pulled out from the bearings. Before the shafts were pulled out and the photos were taken, the shafts were rotated

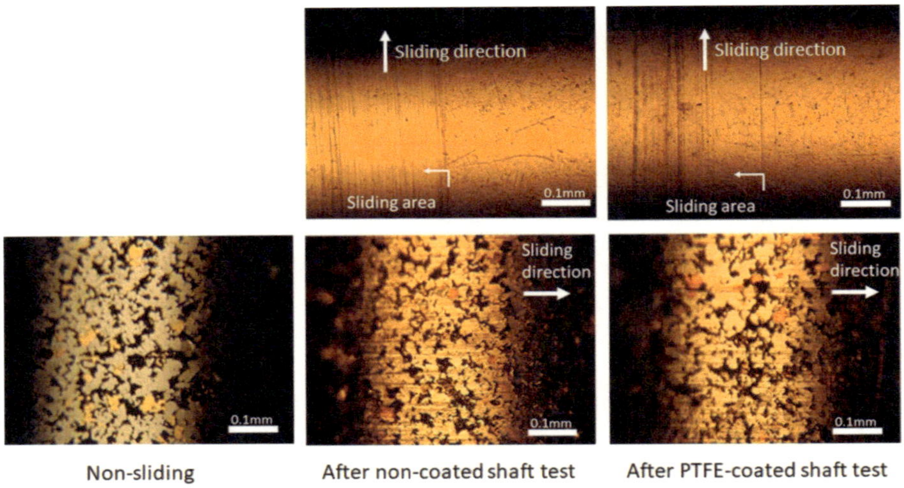

Figure 6.
Shaft and bearing surfaces before and after the tests. Upper: shaft surfaces. Lower: bearing surfaces.

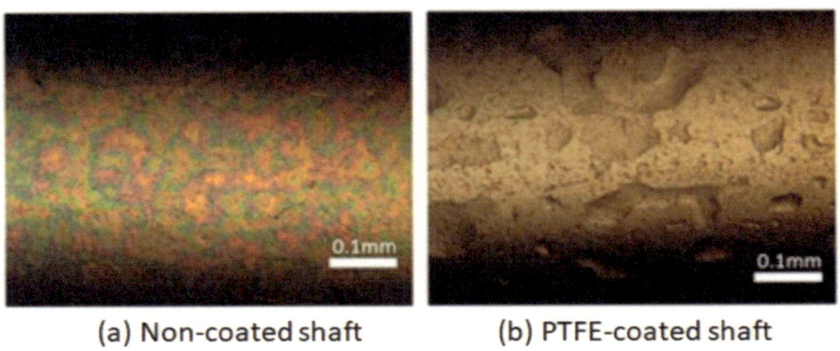

Figure 7.
Oil deposition on the shafts.

several times by hand in the bearings. There were many oil droplets on the PTFE-coated shaft surface, while only interference patterns of a thin film of oil were found on the noncoated shaft surface, which suggested the richer oil amount for the PTFE-coated shaft and bearing pair. However, very thin oil film on the non-coated shaft might not have suggested the less oil amount directly, because it could be due to rapid oil spreading on the shaft before the photos were taken. Next, the oil in the bearing clearances was observed directly to clarify.

The projections of the bearing clearances with fixed shafts were taken with a profile projector (Mitsutoyo PJ-3000F, Japan) and were shown in **Figure 8(a-1)** and **(b-1)**. The configuration of the projector and the tested bearing is shown in **Figure 9**.

The shafts were rotated by hand several times prior to the observation. Then the shafts were rotated about 45 degrees for several seconds from the initial positions. **Figure 8(a-2)** and **(b-2)** shows the photos at that time. Though clear individual figures of oil droplets were lacking, because they were deposited not only in the plane direction but also in the depth direction, there were many and large oil droplets indicated by arrows in the clearance between the PTFE-coated shaft and the bearing, while only a few and small droplets on the noncoated shaft. The difference between (a) and (b) in **Figure 7** must have been caused by this.

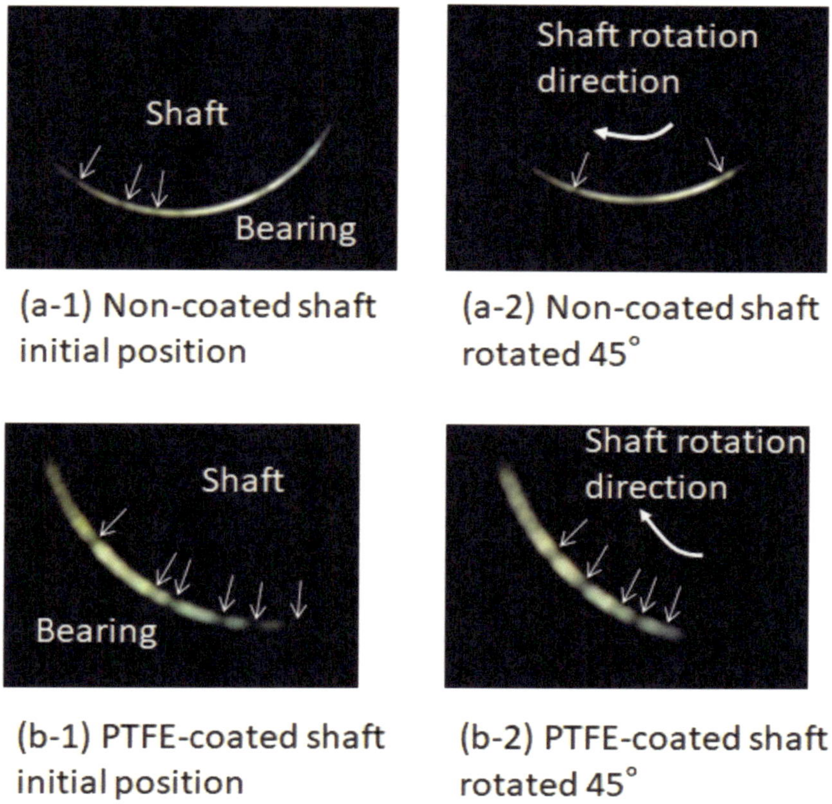

Figure 8.
Bearing clearance projections. Arrows indicate the position of the droplets.

In addition to the large amount of the oil droplets, droplets on the PTFE-coated shaft were observed to move together with shaft rotation, which leads to oil circulation in the clearance, from the moving images taken during the 45° shaft rotation.

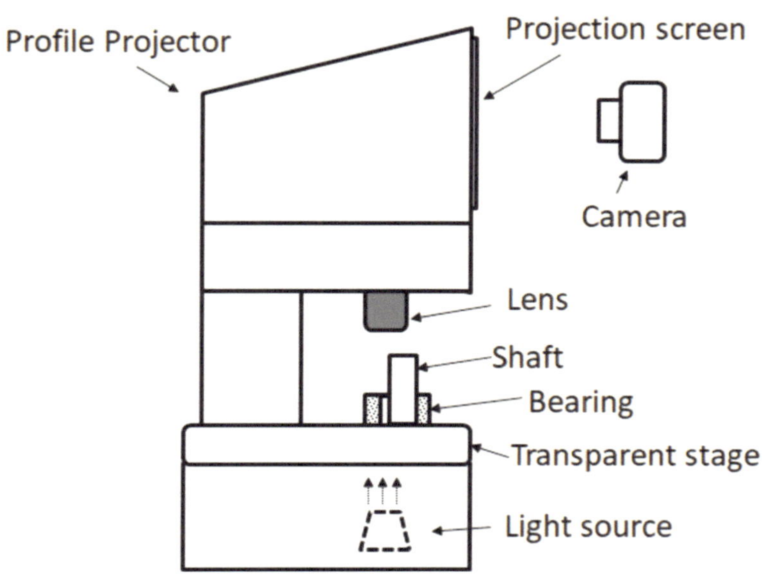

Figure 9.
Configuration of the projector and the tested bearing.

New droplets were observed to emerge from the edge of the clearance and move in the same manner, that is, oil exudation in the clearance for the PTFE-coated shaft was clearly observed. No conspicuous motion of the oil droplets was observed in the clearance for the noncoated shaft during shaft rotation.

To summarize, much larger amount of oil was retained in the bearing clearance for the PTFE-coated shaft (i.e., low wettable shaft) than that for the noncoated shaft (i.e., highly wettable shaft), and oil circulation in the bearing clearance (along with shaft rotation) was observed for the PTFE-coated shaft. This means the oil-rich and secure oil exuding condition for the sliding area of the PTFE-coated shaft and relatively oil-poor condition for the noncoated shaft. Smaller friction of the PTFE-coated shaft was probably attributable to this oil-rich condition.

The past literature has showed that the oil-poor condition increases the friction of sintered bearings [15]. More frequent solid–solid contact could occur for a less amount of oil condition in the sliding area, and thus was naturally one of the reasons for larger friction of the noncoated shaft. The oil-richer condition leads to the lower friction of the PTFE-coated shaft.

4.3 Other effects to cause the difference in friction

In this section, possible mechanisms other than those mentioned in the previous section that caused the difference in friction are discussed.

One of the effects to be noticed especially for poor oil condition is capillary force. Less amount of oil supply to the sliding area would pose synergistic effect of poor lubricity and capillary force on friction increase to the bearing.

Figure 10 shows the diagram of the oil meniscuses in the bearing clearance. Capillary force (F) is generated by negative Laplace pressure in the oil film. Assume that the oil filled the clearance plane symmetrically. Ignoring the effects of hydrodynamic force, the bearing end effect and oil droplets in the rest of the bearing clearance, and approximating zero contact angles between the oil and bearing/shaft

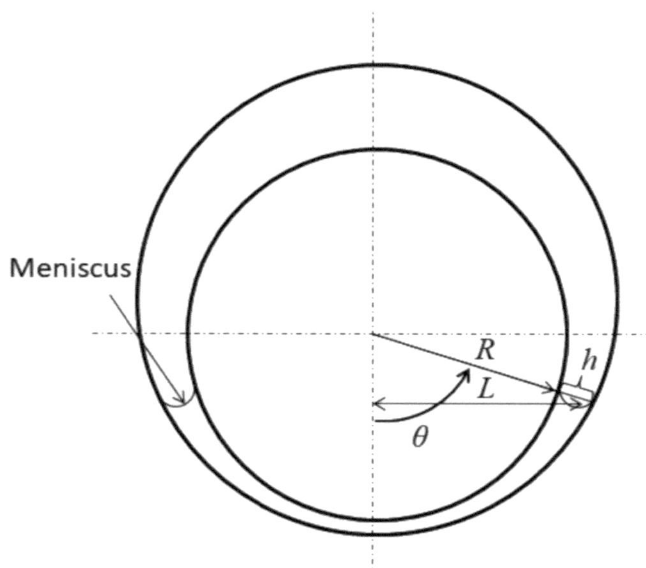

Figure 10.
Diagram of oil meniscuses in bearing clearance.

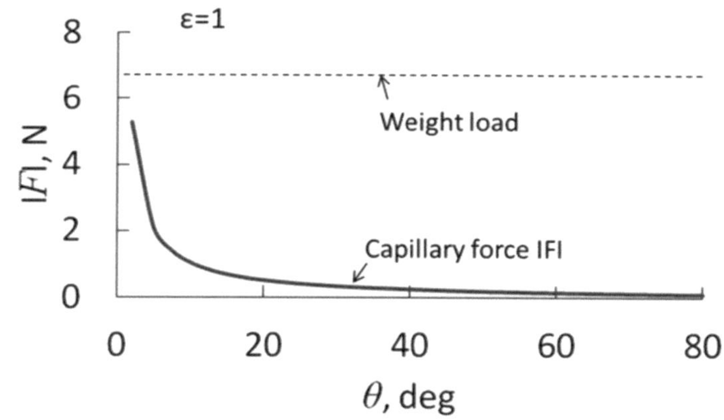

Figure 11.
Calculated capillary force.

surface (the curvature of the meniscus to be −h/2), the capillary force could be estimated as follows:

$$F = 2LB\Delta p = -2LB\gamma/(h/2) \quad (2)$$

$$h \approx c(1 - \varepsilon\cos\theta) \quad (3)$$

$$L = R\sin\theta \quad (4)$$

where, B is the bearing width, Δp is the Laplace pressure, γ is the surface tension of the oil, c is the bearing radial clearance, ε is the eccentricity ratio, and R is the shaft radius.

Figure 11 shows the calculated capillary force along with ε = 1 (which denotes the contact of the shaft and the bearing), c = 6.5 μm, and the tested bearing dimensions. |F| increases with a decrease in θ continuously and increases rapidly at an angle of 10° or less. **Figure 11** indicates that smaller the oil amount is, larger the capillary force becomes and consequently friction becomes larger.

To estimate the effect of the capillary force on the bearing friction, assume the oil filled by θ = ±30° for the noncoated shaft and θ = ±80° for the PTFE-coated shaft in the bearing clearance, respectively. Using Eqs. (2)–(4), F is −0.34 N for θ = ±30° and −0.11 N for θ = ±80°. Thus, the sum of the loads by weight (6.7 N) and capillary force, the actual load, for the PTFE-coated shaft, is 3% less than that of the noncoated shaft. Considering that friction reduction rate was 20–50%, this load difference would not be negligible. In considering a smaller θ, |F| increases rapidly and the actual load for a noncoated bearing becomes large. Hence, the capillary force due to less oil could be the part of the larger friction.

Another possible mechanism of lower friction for the PTFE-coated shaft was slippage between the oil film and the shaft surface. A hydrophobic surface may cause boundary slip—a measure of relative fluid velocity at the solid-fluid interface [9, 16, 17] that leads to low friction under boundary lubrication. This effect was examined by the following friction tests.

The tests were conducted by using a ball-on-plate reciprocating friction tester for PTFE-coated and noncoated plates in the same oil as impregnated oil. **Table 2**

Ball	Material: JIS SUS304 stainless steel Diameter: 10 mm
Plate	Material: JIS SUS304 stainless steel Surface roughness: Ra 0.2 μm
Load	1.6 N
Stroke	30 mm
Sliding speed	30 mm/s

Table 2.
Test conditions and materials for ball-on-plate tests.

indicates the test materials and conditions. JIS SUS 304 stainless steel was used for the test material instead of JIS SUS 420J2 stainless steel (the shaft material) due to its easy preparation. JIS SUS 304 stainless steel was verified in advance to have as large oil wettability as JIS SUS 420J2 stainless steel by measuring the contact angles of the oil. The smallest load (1.6 N) the tester used could make was chosen to prevent the solid-solid contact as much as possible.

Figure 12 shows the time evolution of the friction coefficient of the initial four reciprocations for the PTFE-coated and noncoated plates. A negative friction coefficient denotes the reverse sliding direction. Both shafts showed the friction coefficient of about 0.08 which were comparable to that of noncoated shaft shown in **Figure 5**. This means that the lubrication condition of the ball-on-plate tests was probably about the same as that of the bearing test of the noncoated shaft and the PTFE-coating showed no particular effect like oil-surface slippage on friction under such lubrication condition.

For testing the solid-lubricity of the PTFE-coated shaft under dry condition, "dried" bearing was made by ultrasonic cleaning of oil-impregnated sintered bearing in acetone for more than 10 minutes and tested by the test rig of **Figure 3**. Both noncoated and PTFE-coated shaft indicated the friction coefficient of 0.2–0.3 which means PTFE-coating had little effect on friction reduction under dry condition. From these results, the lubricity of the PTFE coating was presumed not to contribute to the friction reduction in the bearing test.

4.4 Mechanism to retain the oil droplets in the bearing clearance

In this section, the mechanism that caused the difference in oil amount in the bearing clearance will be discussed. Oil droplets in the bearing clearance will be modeled as micro-liquid bridges between slightly tilted plates. This modeling is appropriate because the angle between the shaft and bearing surfaces is usually very small.

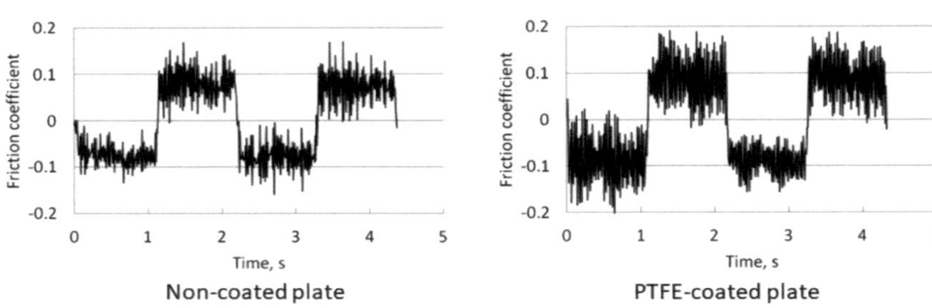

Figure 12.
Friction coefficient of ball-on-plate tests.

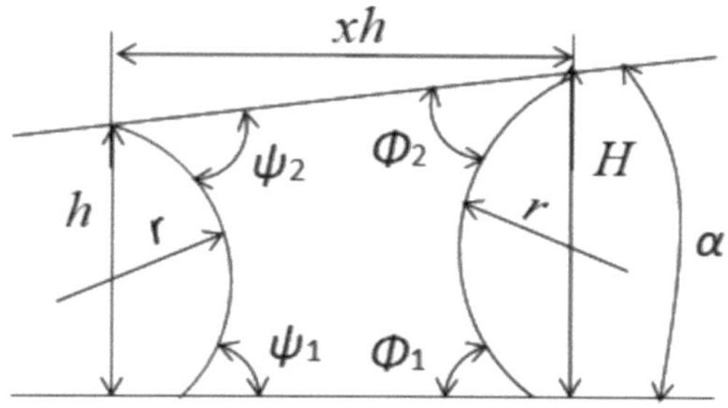

Figure 13.
Liquid bridge between tilted plates.

Figure 13 illustrates the drum-formed liquid bridge between tilted plates. The meniscus radius r in **Figure 13** is expressed by Eq. (5) when the radius of the drum (liquid bridge thickness) is much larger than that of the meniscus and the effect of gravity is ignorable [18].

$$r \approx -\gamma/\Delta p = \text{const.in the meniscus} \tag{5}$$

r is constant for a certain liquid bridge and the following relation holds geometrically:

$$H/h = \{\cos\phi_1 + \cos(\phi_2 + \alpha)\}/\{\cos\psi_1 + \cos(\psi_2 - \alpha)\} = 1 + x\tan\alpha \tag{6}$$

where x is the aspect ratio of the liquid bridge (thickness/height) assuming α is small. To satisfy this relation, contact angles have to vary according to the tilt angle and liquid bridge thickness.

The stability of liquid bridges between plates has been studied by several researchers [19–21]. According to these studies, the liquid bridge between parallel plates with a fixed contact angle between the liquid and the surface becomes unstable (which means a significant change of the liquid bridge in configuration and position) when one of the plates was tilted, even by a small amount [21]. This is because it is impossible to satisfy above relationship for the liquid-bridge with fixed contact angle.

Usually, contact angles could vary between receding and advancing contact angles, which is called contact angle hysteresis [22, 23], and the variable range depends on the liquid and surface material. According to the measuring method described in [22], the advancing angles were measured to be about 15 and 45°, and the receding angles were about 12 and 12.5° for noncoated and PTFE-coated plates, respectively. These results mean that contact angles could be in the following range: $12.5° \leqq \varphi_1, \varphi_2, \psi_1, \psi_2 \leqq 45°$ for the PTFE-coated plate.

Assume that these contact angles could be applied to the liquid bridges. x = 3, $\varphi_1 = \varphi_2 = 12°$, and $\psi_1 = \psi_2 = 15°$ for the noncoated plate combination according to the above measured values, then $\alpha \approx 0.22°$ from Eq. (6). In the same way, for the PTFE-coated plate combination, x = 3, $\varphi_1 = \varphi_2 = 12.5°$, and $\psi_1 = \psi_2 = 45°$, then $\alpha \approx 5.7°$. For larger α, liquid bridges become unstable. To verify this calculation, liquid bridge behavior was observed between plates with a 0.5-mm gap. A photo example of the test configuration and a schematic of the tester were shown in **Figure 14**. The bright line and blue two spots on the liquid bridge were the reflection of background and illumination lamp. The plates are initially set in parallel, and then rotated around the bottom center of the liquid bridge in steps of 0.5°. As shown in the figure, x could be read about 3.

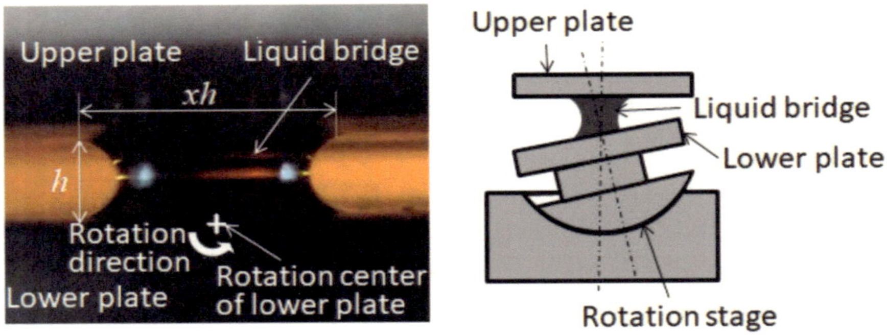

Figure 14.
Example photo of liquid bridge test configuration and schematic of tester.

For the noncoated upper and lower plate combination, the liquid bridge flowed in the narrower gap direction as soon as the tilt started. This means that the liquid bridge became unstable at a tilting angle less than 0.5°. For the PTFE-coated upper and lower plate combination, the liquid bridge flowed at tilting angles of 4~6.5° in several attempts. These results indicate that Eq. (6) is applicable for these conditions though Eq. (6) is not accurate, particularly when liquid bridge thickness is small, because the meniscus curve does not hold an arc shape but a more complex one (e.g., nodoid shape [20]).

We then applied Eq. (6) to the bearing clearance to estimate the possible maximum size of the oil droplet for the position in the clearance. The lower and upper plates can be considered to be the shaft and the bearing, respectively. The bearing surface has wettability similar to that of the noncoated shaft, which was roughly confirmed by oil droplet contact angle measurements. $\varphi_1 = \varphi_2 = 12°$, $\psi_1 = \psi_2 = 15°$ for noncoated shaft and $\varphi_1 = 12°$, $\varphi_2 = 12.5°$, $\psi_1 = 15°$, and $\psi_2 = 45°$ for PTFE-coated shaft, respectively, and $\tan\alpha = dh/d(R\theta)$ were used for the calculation. **Figure 15** shows the calculated critical oil bridge thickness (xh), that is, the largest possible size of oil droplets in the bearing clearance.

From **Figure 15**, the PTFE-coated shaft indicated larger critical oil bridge thicknesses and could retain much larger oil droplets in the bearing clearance than those

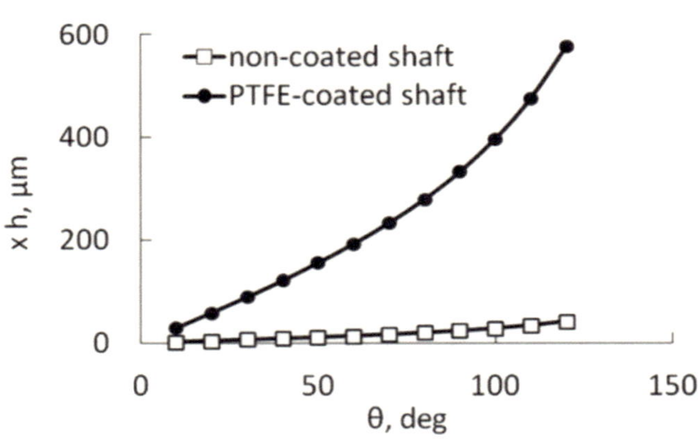

Figure 15.
Critical oil bridge thickness xh.

for the noncoated shaft. This effect may yield the results shown in **Figures 6** and 7 and lead to an oil-rich condition for the PTFE-coated shaft.

In addition to large oil droplets, clear movement of the oil droplets in the bearing clearance with shaft rotation was observed for the PTFE-coated shaft, as described in Section 4.2. This circulative movement in the clearance would induce the oil exudation from the bearing body probably contributing to large oil amount.

To investigate the movement of the oil droplets, the dragging effect of the oil droplets by the PTFE-coated shaft was examined by simple tests shown in **Figure 16**. An oil droplet was bridged between the parallel plates with 1-mm gap, and the upper plate was moved in parallel to the lower plate. The plates were SUS304 stainless steel and the upper plate of **Figure 16(a-1)** and **(a-2)** was PTFE-coated. The upper plates moved to the right by a distance of about 3 mm from the position of **Figure 16(a-1)** and **(b-1)** to the position of **Figure 16(a-2)** and **(b-2)**, respectively.

As can be seen from **Figure 16**, the PTFE-coated plate dragged the oil droplet firmly while the bare stainless plate did a little.

Figure 17 indicates the relationships between the upper plate travel distance and that of oil droplet. The oil droplet appears to stick to the PTFE-coated plate and to slip on the bare plate. This property may have caused the oil droplet circulation in the bearing clearance for the PTFE-coated shaft.

These phenomena would be a matter of energy but could also be explained by contact angle hysteresis. When the plate began to move, the oil droplet had to be deformed to retain the adhesion and the contact angle changed. As allowable contact angle range for the bare plate was much smaller than that of the PTFE-coated

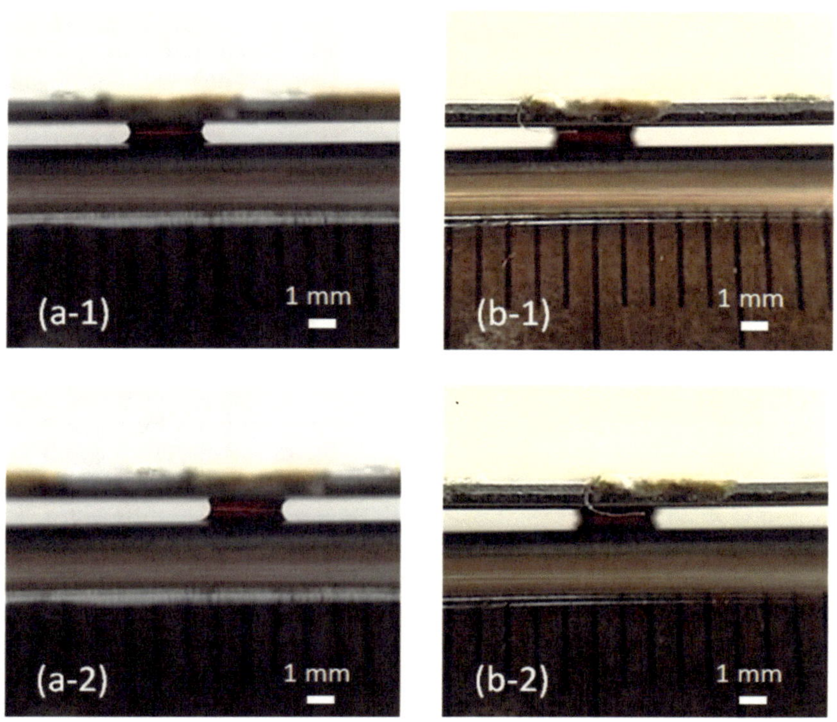

Figure 16.
Photos of oil droplet dragging test. (a-1) PTFE coated upper plate, original position. (a-2) 3 mm moved position. (b-1) Bare plate, original position. (b-2) 3 mm moved position.

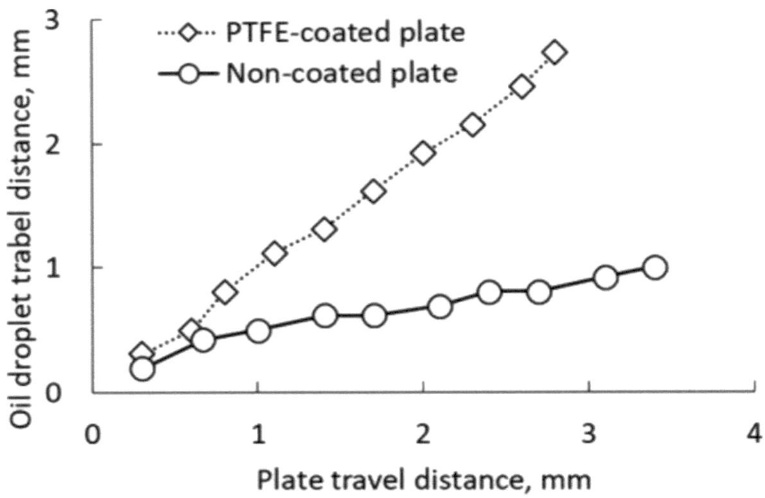

Figure 17.
Relationships between the upper plate travel distance and that of oil droplet.

plate, the interface between the oil droplet and the bare plate became unstable faster and caused to slip.

5. Conclusions

The oil-impregnated sintered bearing with PTFE-coated shaft showed lower friction than that with the noncoated (bare metal) shaft. The PTFE-coated shaft was found to retain a larger amount of oil droplets in the bearing clearance than the noncoated shaft and make secure oil circulation in the clearance. These effects would yield the oil-rich and better lubrication condition leading to lower friction of the PTFE-coated shaft.

The lower friction of the PTFE-coated shaft is attributed to the lower wettability to the impregnated oil than that of the noncoated shaft. Large contact angle hysteresis concomitantly with the low wettability makes it possible to retain larger oil droplets in the bearing clearance. The low wettable surface drags the oil droplets firmer than the high wettable surface, which generates the secure oil circulation in the bearing clearance for the PTFE-coated shaft.

Low wettability is often associated with a large contact angle hysteresis but probably it is not always the case. Therefore, note that probably the results of this study are not applicable to all kinds of low wettable shafts. The large contact angle hysteresis of the PTFE-coated shaft could also be caused by nonuniformity of the PTFE transfer film due to our rough production method [21].

However, generally speaking, using low wettable shafts must be one of the effective means to improve the lubrication condition and reduce the friction of oil-impregnated sintered bearings, though PTFE transfer film coating, which we used as low wettable coating in this study, is not wear-resistant and not a proper material for practical uses.

Acknowledgements

I would like to thank Fumiaki Katagiri and Yuta Kawabe, the former students of Institute of Technologists, for their devoted effort to the experiments.

Notes

Most of the content of this chapter is based on Ref. [5].

Author details

Naofumi Hiraoka
Institute of Technologists, Gyoda, Saitama, Japan

*Address all correspondence to: hiraoka@iot.ac.jp

References

[1] Watanabe T. Present status of powder metallurgy in Japan. The International Journal of Powder Metallurgy & Powder Technology. 1980;**16**(3):249-253. (in Japanese)

[2] Watanabe T. Porous sintered bearings. Journal of the Japan Society of Powder and Powder Metallurgy. 2001;**48**(9):769-776. (in Japanese). DOI: 10.2497/jjspm.48.769

[3] Porite Corporation Product Information [Internet]. Available from: http://www.porite.co.jp/product/example01.html [Accessed: 30-07-2018]

[4] Tanabe S. Energy saving tribology in impregnated sintered bearings. Journal of Japanese Society of Tribologists. 2006;**51**(5):353-358. (in Japanese)

[5] Hiraoka N, Katagiri F, Kawabe Y. Effects of shaft surface wettability on the friction of oil-impregnated sintered bearings—Effect of contact angle hysteresis. Tribology Online. 2016;**11**(2):403-409. DOI: 10.2474/trol.11.403

[6] Kaneko S. Porous oil bearings. Journal of Japanese Society of Tribologists. 1993;**38**(9):773-778. (in Japanese)

[7] Quan Y-X, Ma J, Tian Y-G, Zhou G-R, Shi G-Y, Wang G-L. Investigation of sintered bronze bearings under high-speed conditions. Tribology International. 1985;**18**(2):75-80. DOI: 10.1016/0301-679X(85)90033-7

[8] Sauer CH, Schatt W, Friedrich E. High performance Cu-basis PM bearings. In: PM into the 1990s. International Conference on Powder Metallurgy. 2. London. 1990. pp. 422-425

[9] Choo JH, Glovnea RP, Forrest AK, Spikes HA. A low friction bearing based on liquid slip at the wall. Journal of Tribology. 2007;**129**(3):611-620. DOI: 10.1115/1.2736704

[10] Ohmori H, Katahira K, Mizutani M. Functionalization of stainless steel surface through mirror-quality finish grinding. CIRP Annals. 2008;**57**(1):545-549. DOI: 10.1016/j.cirp.2008.03.131

[11] Jap Soc Tribol. Tribology Handbook. Tokyo: Yokendo; 2001. pp. 744-747. (in Japanese)

[12] Tanaka K, Miyata T. Studies on the friction and transfer of semicrystalline polymers. Wear. 1977;**41**(2):383-398

[13] Wheeler DR. The transfer of polytetrafluoroethylene studied by X-ray photoelectron spectroscopy. Wear. 1981;**66**(3):355-365. DOI: 10.1016/0043-1648(81)90128-9

[14] Lauer JL, Bunting G, Jones WRJR. Investigation of frictional transfer films of PTFE by infrared emission spectroscopy and phase-locked ellipsometry. Tribology Transactions. 1988;**31**(2):282-288. DOI: 10.1080/10402008808981824

[15] Raman R, Babu LV. Tests on sintered bearings with reduced oil contents. Wear. 1984;**95**:263-269. DOI: 10.1016/0043-1648(84)90141-8

[16] Leger L, Hervert H, Pit R. Friction and flow with slip at fluid-solid interfaces. ACS Symposium Series; **2001**:154-167

[17] Voronov RS, Papavassiliou DV. Review of fluid slip over superhydrophobic surfaces and its dependence on the contact angle. Industrial and Engineering Chemistry Research. 2008;**47**:2455-2477. DOI: 10.1021/ie0712941

[18] Taura H, Kaneko S. Meniscus forces of liquid bridge between two

parallel planes (comparison between exact solution and approximations). Transactions of the JSME. 2012;**78**(790):331-342. (in Japanese). DOI: 10.1299/kikaic.78.2266

[19] Chen T-Y, Tsamopoulos JA, Good RJ. Capillary bridges between parallel and non-parallel surfaces and their stability. Journal of Colloid and Interface Science. 1992;**151**(1):49-69. DOI: 10.1016/0021-9797(92)90237-G

[20] Langbein D. Stability of liquid bridges between parallel plates. Microgravity Science and Technology. 1992;**5**(1):2-11

[21] Concus P, Finn R. Discontinuous behavior of liquids between parallel and tilted plates. Physics of Fluids. 1998;**10**(1):39-43. DOI: 10.1063/1.869547

[22] Gennes P-G, Brochard-Wyart F, Quere D. Gouttes, Bulles, Perles et Ondes. Kyoto: Yoshioka Shoten; 2012. pp. 67-68. (Japanese Edition)

[23] Adamson AW, Gast AP. Physical Chemistry of Surfaces. New York: John Wiley & Sons; 1997. pp. 355-362

Chapter 3

Understanding Wettability through Zeta Potential Measurements

Syed Taqvi and Ghada Bassioni

Abstract

Traditional wettability measurement practices, introduced as early as the late 1950s, are still perceived as reliable industrial methods for characterizing wettability. These techniques, in contrast to the contact angle method, provide indices that can quantitatively describe the degree of wetting state. Nevertheless, these approaches can determine the current state of wettability, subject to limitations. By employing the zeta potential measurements technique, a complete wettability profile is derived. In this chapter, the theory behind this technique is discussed. A case study is presented where the wettability of limestone rock is investigated, as crude oil and asphaltenic solutions of varying concentration are added to the limestone-water suspension. Findings of this study are discussed with the possible mechanism in effect when wettability alteration is observed.

Keywords: zeta potential, water-wet, oil-wet interfaces, calcium carbonate, asphaltene

1. Introduction

In the oil and gas industry, wettability studies are extensively carried out on reservoir rocks as it is one of the primary factors that control location, saturation distribution, and flow behavior of reservoir fluids [1]. Consequently, it influences several petrophysical properties such as capillary pressure, relative permeability, waterflood behavior, electrical properties, and enhanced oil recovery (EOR) [1–3].

Traditional methods, devised by researchers over time, have classified wettability of rocks into three main states: (i) oil-wet, (ii) water-wet, and (iii) neutral wet. Oil-wet refers to the state of the reservoir rock when oil, in presence of other fluids, preferentially covers the rock surface. As a consequence, oil forms a thin layer on the surface of the rock, while other fluids remain in the center, as seen in **Figure 1(a)**. For the water-wet scenario, water in the presence of other fluids (i.e., oil) preferentially covers the rock surface. As a result, water forms a thin layer on the surface of the rock, while oil remains in the center, as seen in **Figure 1(b)**. In the case of neutral wetting, the reservoir rock surface does not show any preferential covering for any fluid. It is not necessary that a reservoir rock strongly adheres to either one of these phenomena; rather cases of heterogeneity may exist where many other states of wettability may be defined, such as mixed wettability, fractional wettability, "Dalmatian" wetting, and speckled wetting [3].

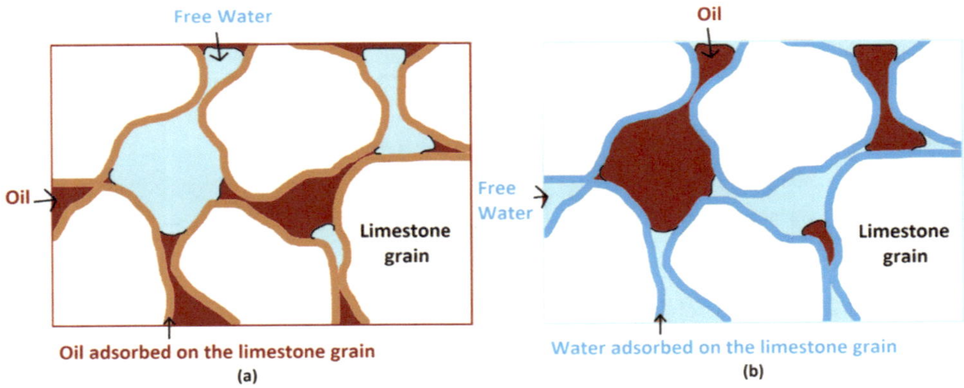

Figure 1.
Schematic diagrams of (a) oil-wet and (b) water-wet states of wettability on a microscopic level.

In the case of mixed wettability, some parts of the rock surface are water-wet, while others are oil-wet [4]. For the fractional wettability scenario, a fraction of the internal rock surface area is wetted by a fluid, either oil or water [5]. "Dalmatian" is a type of nonuniform wettability where oil-wet and water-wet regions, smaller than

		Wettability measurement technique	Advantages	Disadvantages
Quantitative	1	Contact angle	Accurate Simple	Surface heterogeneity can affect reading Cannot provide information regarding organic coatings Impurities can affect reading Liquid drop can form many stable contact angles
	2	Amott/ Amott-Harvey	Reliable	Not very effective to measure neutral wettability No validity as an absolute measurement
	3	United States Bureau of Mines (USBM)	Reliable	No validity as an absolute measurement
Qualitative	4	Imbibition rate	Quick Simple	Not very accurate Imbibition rate does not solely depend on wettability but also other factors
	5	Microscopic examination	In situ wettability alteration can be observed	Based on affinity with no degree of wettability measured
	6	Flotation	Quick	Sensitive to only strongly wetted systems
	7	Nuclear magnetic resonance (NMR)	Ability to study fractional wettability	Requires complex and expensive machinery
	8	Dye adsorption	Quick Simple Cost effective Ability to study fractional wettability	Not very accurate

Table 1.
Summary of wettability measurement techniques.

Index	Oil-wet	Neutral wet	Water-wet
Amott wettability index water ratio (IW)	0	0	>0
Amott wettability index oil ratio (IO)	>0	0	0
Amott wettability index (WIA)	−1.0 to −0.3	−0.3 to 0.3	0.3 to 1.0
United States Bureau of Mines (USBM)	−1	0	1
Contact angle (θ)	>90°	90°	<90°

Table 2.
Index/measurement values depicting different wetting states for different quantitative techniques.

the size of a pore, are mixed randomly [6]. Speckled wetting refers to the situation where oil tends to be trapped in pore throats rather than the pore bodies [7].

All the developed techniques, currently employed, express wettability either quantitatively or qualitatively. A common limitation of all these traditional methods is their inability of in situ measurement of the state of wettability while the wettability alteration process takes place [8, 9]. **Table 1** summarizes the other specific advantages and limitations of wettability measurement practices that are commonly used. For the quantitative wettability techniques, **Table 2** shows the index/measurement values for the corresponding wetting states.

2. Zeta potential approach

Zeta potential is defined as the difference in potential between the particle and its ionic atmosphere surrounding medium, measured in the plane of shear, as depicted in **Figure 2** [10]. In the ideal case, the potential at the Stern layer, Ψ_d, is desired potential. However, due to the non-conductive nature of most minerals, it becomes challenging to measure it [11]. Thus, zeta potential measurements are

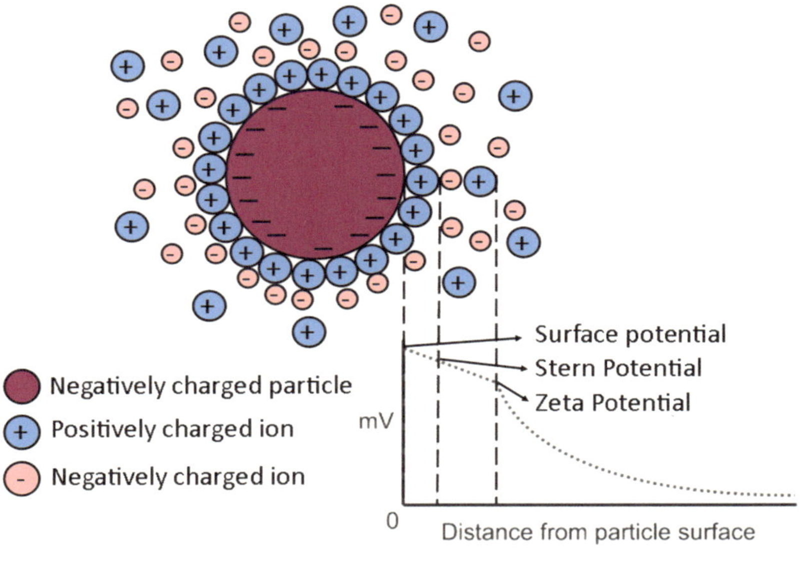

Figure 2.
Diagram illustrating the surface potential, stern potential, and zeta potential as a function of distance of a negatively charged particle in a dispersion medium.

considered that are slightly greater than the Stern potential [10]. Mechanistically speaking, when a solid particle is dissolved in water, the charge on its surface leads to the formation of counterions [12]. For example, in the case of a limestone slurry, the co- and counterions, Ca^{2+} and OH^{-}, are formed. The charged particles then attract a layer of counterions from the aqueous phase, known as the Stern layer.

Due to ionic radius considerations, the strongly adsorbed anions do not fully offset the surface charge, and hence, a second layer of more loosely held counterions forms, known as the diffuse layer. This diffuse layer is made of free ions which are mobile under the influence of electric attraction and thermal motion. The two layers combined are known as the electric double layer (EDL). At a certain distance away from the charged particle, the surface charge will be fully balanced by the counterions, and beyond this point, a bulk suspension with a balance of positive and negative electrolyte exists [9, 12]. Particles, with a high surface charge, form a large double layer, which prevents particles from getting close to each other because of the electrostatic repulsion between them due to identical charges. However, this behavior aids in stabilizing suspensions [13].

Due to the presence of electric charges in an applied electric field, different phenomena occur in the system, which are collectively known as electrokinetic effects. The zeta potential can be measured based on different phenomena classified as electrokinetic or electroacoustic. Electroacoustic differs from electrokinetic as the former uses ultrasound waves to induce movement of particles in one direction. Unlike the latter, it does not utilize an electric field. However, they are both used to measure the surface charges of stable suspensions composed of small particles of sizes less than 10 μm.

2.1 Electrokinetic measurement

The electrokinetic phenomena are differentiated based on the induced motion of particles [14]. These techniques include:

1. Electrophoresis

2. Electroosmosis

3. Streaming potential

4. Sedimentation potential

2.1.1 Electrophoresis

Electrophoresis is considered to be the most widely used technique in electrokinetics to measure the zeta potential. It considers the movement of a charged particle relative to its liquid medium under the influence of the electric field [14]. It occurs when an electric field is applied across the sample electrolyte and causes the suspended particles to move [14]. However, with an opposing viscous force, equilibrium is achieved. As a result, the particles move with constant velocity. This velocity, or electrophoretic mobility, depends on [14]:

- The strength of the electric field

- The dielectric constant of the liquid

- The medium's viscosity
- The zeta potential

The electrophoretic mobility is measured in a microelectrophoresis system. This system consists of a cell with electrodes at both ends applying a potential across the sample. The suspended particles will move toward the opposite-charged electrode during which the velocity is measured.

2.1.2 Electroosmosis

Electroosmosis, on the other hand, is associated with the movement of the liquid medium relative to a stationary charged surface. The walls of the capillary cell carry a surface charge which causes the liquid adjacent to the wall to flow, once the electric field is applied. This electroosmotic flow superimposes the colloidal particles' electrophoretic mobility [14]. In a closed system, this flow is countered by another reverse flow down the center of the capillary cell. Due to that, there exists a point, known as the stationary layer, in which the two flows cancel each other and result in a zero net electroosmotic flow [14]. At that point, the true electrophoretic velocity is measured [14]. The position of the stationary layer depends on the geometry of the cell.

2.1.3 Streaming potential

Streaming potential phenomenon is associated with the generation of an electric field when a liquid is forced to flow past a stationary charged surface [14]. The potential difference created is known as the streaming potential. This technique is used to measure the zeta potential for large particles in unstable suspensions.

2.1.4 Sedimentation potential

Sedimentation potential, lastly, deals with the generation of an electric field due to the movement of charged particles relative to a stationary medium, in gravitational or sedimentation fields in centrifuges [15]. It is considered to be the opposite of electrophoresis. However, it is rarely used due to its difficulty in measuring the electrophoretic velocity [15].

2.2 Electroacoustic

In electroacoustics, the motion is due to the different densities of particle and medium [16]. This motion disturbs the double layer around the particles and eventually forms a dipole moment. The summation of the dipoles results in an electric field, known as colloidal vibration current (CVI). A two-electrode sensor detects this current and uses it to compute the zeta potential.

The measurement depends on several factors such as zeta potential, density difference between the medium and suspending particles, viscosity, dielectric permittivity of the liquid, and particles' weight fraction. This technique has several advantages over the electrophoresis which are [16]:

- Measurement of high solid content systems, up to 50% volume fraction, with no need for dilution
- Less sensitive to contamination

- Higher precision (±0.1 mV)
- Low surface charges (down to 0.1 mV)
- Accurate for nonaqueous dispersions

2.3 Significance to the wettability study

Zeta potential measurements are used in nearly all industries that involve colloidal systems. In addition, measuring zeta potential is a reflection of characterizing the surface charge of a suspended particle [17]. Recently, zeta potential measurements have been used extensively to study various biomaterials such as proteins, leukocytes, and DNA [18–20]. Moreover, they have been used to understand the behavior of nanoparticles with human cells, beneficial for applications in the field of medicine [21]. Adhesive properties of various polymers and biomaterial have been investigated using the zeta potential approach [20, 22]. In a study, conducted by Bassioni [23] on environment-friendly construction, zeta potential measurements were used to identify the presence of anionic superplasticizers on cement particles. In addition, Bassioni and Ali [24] performed zeta potential measurements to identify effective dosage of additives to oil-well cement.

In another study, conducted by Bassioni [25], $CaCO_3$ showed potential to adsorb significant quantities of anions due to its positive surface charge. Moreover, zeta potential measurements were made to determine the surface charge of the calcium carbonate particles in the presence of varying dosages of the adsorbates [25]. Since wettability refers to the adsorption of a fluid, in presence of immiscible fluids, onto a rock surface, zeta potential measurements are a promising way to measure wettability. As compared to all other wettability measurement techniques discussed, the zeta potential approach is significant as it can study the wettability alteration in situ [26]. Wettability studies can be carried out on the reservoir rock in order to determine its tolerance and its interaction with inhibitors, prior to any injections to the reservoir. Predictions can be made to study the effect of reservoir conditions onto the rock wettability.

Several studies have been conducted where researchers have employed the concept of zeta potential for different applications. Elimelech et al. used zeta potential measurements to study reverse osmosis membranes [27]. In another study, zeta potential was used to determine the optimum dosage of additives to cement for hydration [28]. Bousse et al. conducted zeta potential measurements on metallic oxide thin films [29]. In addition, another study carried out zeta potential measurements on some transition metal oxides at high temperature [30].

3. Case study

3.1 Methodology

Several experiments were conducted in order to study wettability on reservoir rocks using zeta potential measurements. These include the wettability study of:

- Crude oil on calcium carbonate in limestone-water mixture
- Water on calcium carbonate in limestone-oil mixture
- Crude oil A on calcium carbonate in limestone-water mixture with inhibitors

- Asphaltenic solutions on calcium carbonate in limestone-water mixture
- Maltenes on calcium carbonate in limestone-water mixture

The experimental setup, as shown in **Figure 3**, mainly comprises of a zeta probe, an electronic stirrer and a mixing cell. Prior to the experiment, the particle size of the powdered reservoir rock (i.e., limestone) was determined using a WJL laser granulometer, as it was suspended in water. For each of the experiment, a colloidal mixture was prepared. For the first study, 200 g of limestone was poured into 200 g of water over a period of 1 minute. The mixture was allowed to sit for 1 minute and then stirred vigorously for 2 minutes before transferring to the mixing cell. In the mixing cells, the mixture is continuously stirred using an electronic stirrer at room conditions. All experiments conducted in this study were carried out using distilled water since wettability effects become very important when brine saturation is lowered [31].The zeta probe was placed and the zeta potential was measured. After the initial zeta potential value was recorded, the fluid of interest is added mL-wise to the mixture, and successive zeta potential measurements were made.

Model DT-1200 Electroacoustic Spectrometer (Dispersion Technology Inc., New York, USA) was used to measure the zeta potential values of the system. Moreover, the device has a zeta potential measurement accuracy of ± 0.1 mV + 0.5, as stated by the manufacturer. Densities of the liquid medium and the suspended solid medium and the suspended solid particle size were the parameter input to the system. The probe was placed in the external cell, and the zeta potential was measured, as illustrated in **Figure 3**. For the proposed zeta potential setup, the zeta probe takes measurements when immersed in a colloidal system. Moreover, the system of solid fluid needs to be smooth so a steady colloidal system exists throughout the analysis. The system works with small particles of sizes less than 10 μm. If aggregation in the system or deposition on the probe occurs, it may result in inaccurate readings.

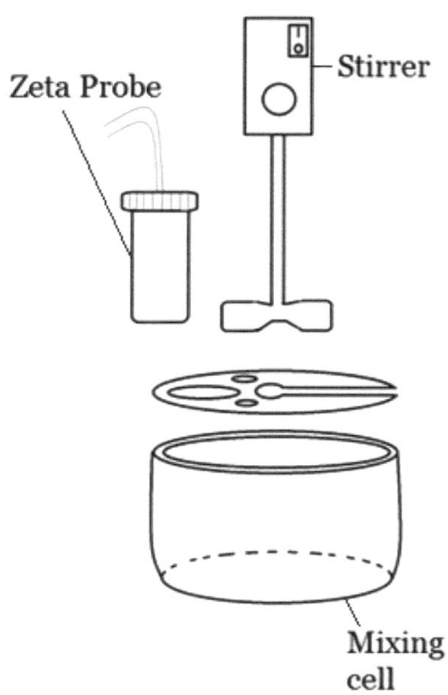

Figure 3.
Experimental setup [26].

3.2 Results and discussion

This section states the results obtained from each analysis and experiment. In addition, it discusses each observation in detail, with possible explanations for each of the finding.

The D_{50} size of the limestone particle was found to be 7.06 ± 0.211 μm with a surface area of 8.628 m^2/g and a density of 2.66 g/cm^3. On the other hand, the PCD analysis showed that both inhibitors, SD and KT, carry a surface charge of −2097 and −1145 mV, respectively. Additionally, the moisture content in each inhibitor was found to be 96.2 and 71.3%, respectively. The ICP-MS analysis, carried out on the SD inhibitor samples, showed significant concentrations of Rb, Ca, and Mg, in descending order. On the other hand, Ni, Al, and Cr were found in significant concentrations in the KT inhibitor (from highest to lowest, respectively).

Various wettability studies were carried out for different crude oil samples on limestone, in presence and absence of inhibitors, as well as of oil derivatives, such as maltene and asphaltenes. The following sections state the findings of each and discuss the results in detail.

3.2.1 Wettability of calcium carbonate for crude oil in limestone-water mixture

Figure 4 shows the results of the zeta potential measurements of the limestone-water suspension as different crude oils are added to the system.

Initially, the surface charge for aqueous limestone suspension was found to be about +30 mV. At this stage, the $CaCO_3$ is completely wetted by water since no crude oil is added yet. Upon addition of crude oil, the zeta potential of the aqueous limestone suspension starts to decrease steadily until a certain limit beyond which a steep fall is observed. This limit differs for each type of oil, as seen from **Figure 4**.

As oil comprises of negatively charged particles [3, 32], oil droplets tend to adsorb to the positively charged $CaCO_3$ surface as a result of electrostatic interaction, displacing the water particles to the bulk. **Figure 5** depicts this competitive adsorption behavior. Once the entire surface is completely wet with the oil droplets, the wettability curve experiences a steep decrease in the zeta potential value, signifying a complete oil-wet surface. The curve flattens out to indicate the end of the adsorption process.

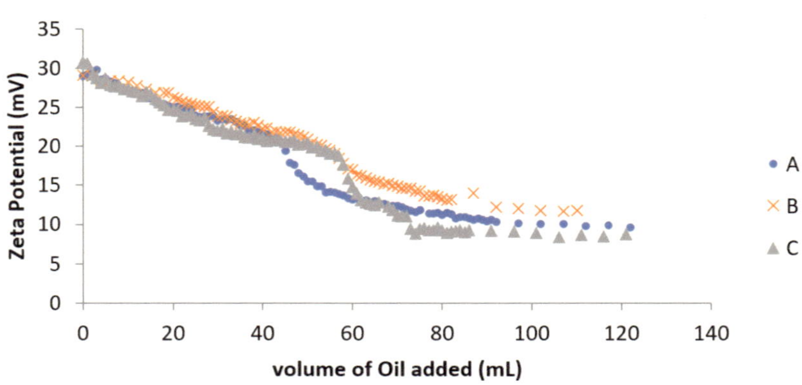

Figure 4.
Zeta potential measurements for aqueous limestone suspension with volume of crude oil added [26].

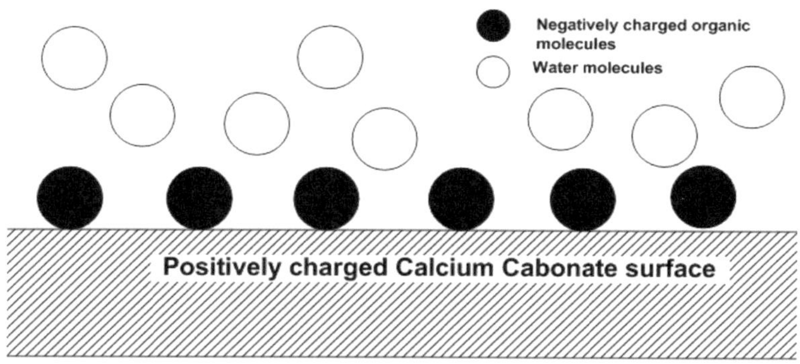

Figure 5.
Negatively charged crude oil particles adsorbing onto positively charged calcium carbonate surface [33].

It can be seen from **Figure 4** that all crude oil samples experience a similar trend upon the addition of crude oil. However, crude oil C tends to have a slightly different trend. This behavior will be explained in the section discussing the wettability alteration due to asphaltene.

As seen from **Figure 4**, it is observed that there exists a general trend in the wettability curve for all crude oil samples. However, the curve appears to be shifted vertically. In accord with our own investigation [26], a previously reported study showed C_7 asphaltene particles, in presence of resins, are negatively charged [34]. Therefore, the more the asphaltene content present, the lower the zeta potential decrease observed. Hence, the vertical shift is observed.

3.2.2 Wettability of calcium carbonate for water in limestone-oil A mixture

Figure 6 shows the zeta potential measurements of the limestone-crude oil mixture as water is added.

The initial zeta potential of the limestone-oil suspension was found to be −15 mV. This signifies that oil carries components with high negative charge, as compared to limestone, lowering the zeta potential value of the limestone suspension to a negative value. Upon addition of water, the zeta potential fluctuates around the average value, as exhibited in **Figure 6**.

Initially, the limestone surface is completely coated by oil (i.e., oil-wet), and the mixture is being stirred smoothly. When water is added to the system, it washes

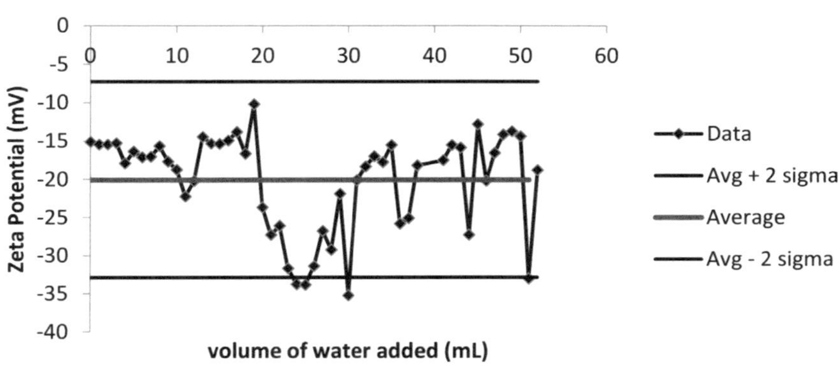

Figure 6.
Zeta potential measurements for limestone-oil A suspension with volume of water added [26].

away the negatively charged adsorbed oil droplets from the surface into the bulk. However, due to electrostatic interactions, these oil droplets tend to return to the limestone surface and adsorb onto it, with time. This continuous behavior results in fluctuations in the zeta potential values. **Figure 7** shows the remainder of the limestone-oil water system when the experiment was halted.

It can be seen in **Figure 7** that deposition of heavy fractions of oil on the limestone surface occurred, forming large chunks of solid matter. The formation of solid chunks became an obstacle for the stirrer in mixing, and the experiment was halted. This situation arose due the addition of water. Water tends to alter the composition of the system, destabilizing the limestone-oil suspension. Due to such disturbance, the heavy fractions aggregate and deposit onto the surface of limestone, as seen from **Figure 7**.

3.2.3 Wettability of calcium carbonate for crude oil A in limestone-water mixture with inhibitors

The wettability experiment was carried out on the aqueous limestone suspension in presence of inhibitors, and the data was compared to the wettability profile of crude oil A in absence of inhibitors, as seen in **Figure 8**.

The initial zeta potential value for the limestone-water inhibitor system for both inhibitors, KT and SD, was 16 mV even though KT has a particle charge of −1145 mV, whereas SD has a particle charge of −2097 mV. The zeta potential decreases with the addition of crude oil to the system. Moreover, both inhibitors decrease the zeta potential of the system to a value of about 4 mV, after 80 mL of crude oil was added. Yet, inhibitor SD experiences an earlier decrease in zeta potential value relative to inhibitor KT.

As compared to the wettability profile of calcium carbonate without inhibitor, it is observed that inhibitor SD decreases the system tolerance to the addition of crude oil, whereas inhibitor KT slightly increases it. In other words, the wettability alteration process, from water-wet to oil-wet, is accelerated by inhibitor SD, while inhibitor KT decreases it. This analysis shows that inhibitor SD is ineffective for crude oil A.

Figure 9 shows an illustration, showing the possible mechanism that the inhibitor particles follow, in the limestone-water inhibitor system, as crude oil is being added.

Figure 7.
Limestone-oil suspension upon addition of 53 mL of water [26].

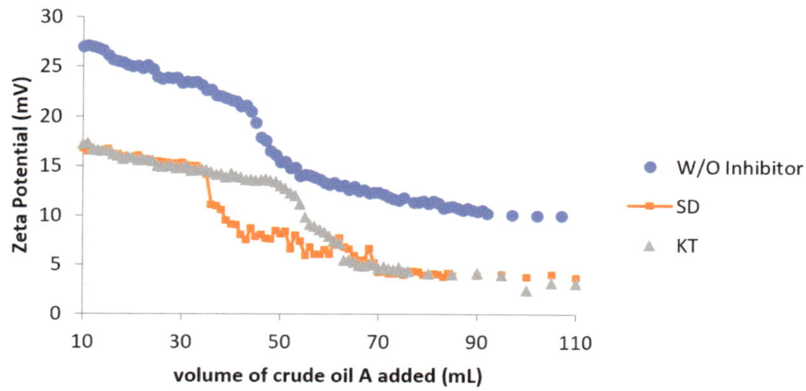

Figure 8.
Zeta potential measurements of aqueous limestone suspension without inhibitor and with inhibitors, KT and SD [26].

In the limestone-water inhibitor system, the inhibitor gets attracted and adsorbed onto the limestone surface due to its high negative particle charge. Since the inhibitor possesses a higher negative charge than the crude oil, upon addition of crude oil, it remains adsorbed on the limestone surface and tends to resist the wettability alteration, keeping the oil particles in the bulk.

The transition is resisted due to the negative charges present on both, crude oil and the inhibitor. These negative charges exert forces of repulsion, preventing the crude oil particles from coming in contact with the limestone surface and keeping them afloat in the oil. By extension, this prevents the deposition of heavy fractions of oil, maintaining the composition of the crude oil. However, it can be seen that in **Figure 7**, the zeta potential value of the suspension decreases to an even lower value upon addition of crude oil. Eventually, the inhibitor particles are forced to desorb due to the increasing concentration of crude oil. Hence, the wettability changes.

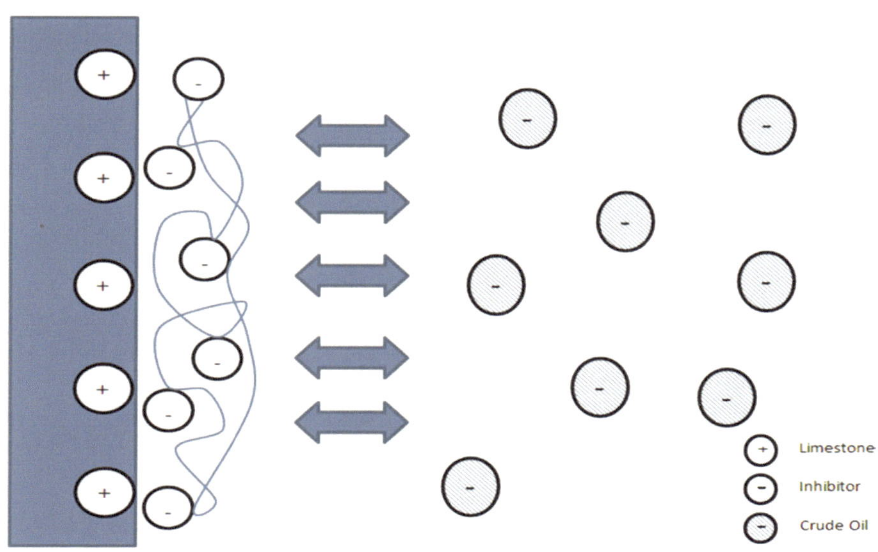

Figure 9.
Mechanism of inhibitor adsorbing on the limestone surface in the presence of crude oil [26].

3.2.4 Wettability of asphaltenic solutions on calcium carbonate in limestone-water mixture

Asphaltene was extracted from crude oil A, and different asphaltene-toluene solutions of varying concentrations were prepared. **Figure 10** shows the wettability curve for each solution.

Initially, the limestone surface in the limestone-water system is completely water-wet. Upon addition of asphaltenic solution, the zeta potential value decreases for all concentrations of asphaltene. This signifies the negative character of the asphaltene molecules in toluene as well as in oil, as supported by Bassioni and Taqvi [26]. As the concentration of the asphaltenic solution increases, the zeta potential tends to fall to a more negative value. All concentrations except 5 wt% follow a similar trend in the decrease of the zeta potential value, upon adding the asphaltenic solution. The reason behind the change in trend in case of 5 wt% asphaltenic solution will be explained in Section 3.3, discussing the wettability alteration due to asphaltene.

Initially, the limestone surface in the limestone-water system is completely water-wet. Upon addition of the asphaltenic solution, asphaltene particles tend to adsorb onto the limestone surface and wet it. The wettability alteration is complete once the wettability curve flattens out, and the plateau, at the end of the wettability curve, marks a completely asphaltene-wet limestone surface.

3.2.5 Wettability of maltenes on calcium carbonate in limestone-water mixture

In addition to the asphaltenes, a wettability study was carried out on the maltenes extracted from crude oil. **Figure 11** shows the wettability curve for maltenes, asphaltenes, and the crude oil.

Initially, the limestone surface in the limestone-water system is completely water-wet. When maltene is added to the system, the zeta potential value decreases until the curve flattens out, as seen in **Figure 11**. Resins, in the maltenes, adsorb onto the surface, and the wettability curve flattens out when the limestone surface is completely maltene-wet.

When compared to the wettability of limestone for crude oil A, the zeta potential values of the maltene wettability curve seem to coincide with it. Moreover, the maltene wettability curve attains a plateau at a zeta potential +20 mV,

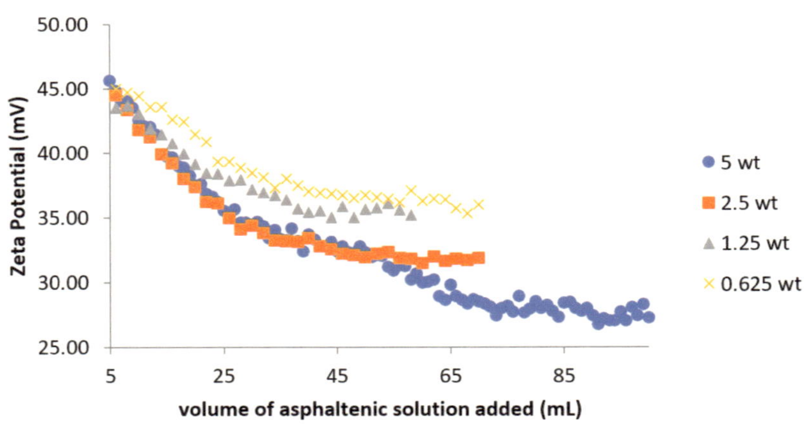

Figure 10.
Wettability of aqueous limestone mixture for asphaltenic solutions of varying concentrations of crude oil A [26].

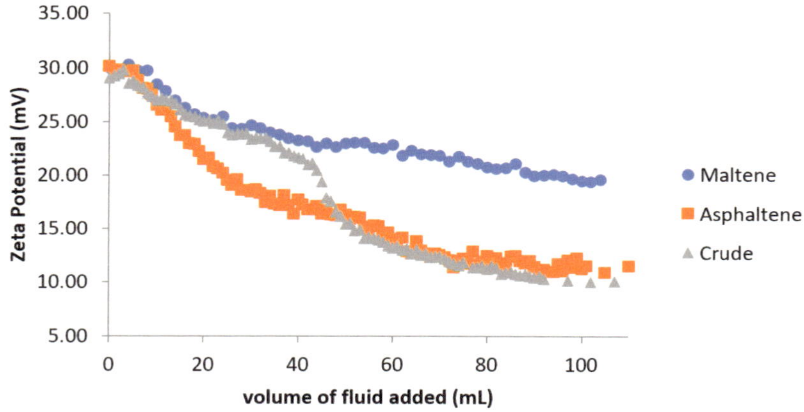

Figure 11.
Wettability profile of limestone for maltene, asphaltene, and crude oil [33].

corresponding to the wettability alteration in the wettability curve of limestone for crude oil. On the other hand, the plateau of the wettability curve for asphaltenic solution coincides with the plateau of the wettability curve for crude oil A. In addition, both wettability curves seem to coincide at a similar value of about +12 mV.

The behavior can be explained such that when crude oil is added to limestone-water mixture, the maltenes, in the crude oil, readily adsorb onto the surface of the limestone particle. After the surface is completely maltene-wet, asphaltenes start adsorbing onto the surface of calcium carbonate. The wettability curve approaches a plateau when the limestone surface is completely asphaltene-wet.

3.2.6 Summary

In contrast to the quantitative techniques discussed in **Table 1**, the zeta potential technique presents an entire wettability profile for the reservoir rock. Wettability measurements are unique for each system. Cases have been presented where different wetting states can be understood from the proposed zeta potential technique. As stated earlier, the quantitative measurement methodologies stated here cannot determine wettability in situ. Thus, if samples of the mixture are drawn at different wetting states, respective values would be observed, as shown in **Table 2**. For the study presented in Section 3.2.1, if a sample from the mixture was obtained anywhere from the beginning of the experiment until where the plateau's first steep drop was observed in **Figure 4** (i.e., ~44 mL of crude oil for oil A), the contact angle method would yield θ values of less than 90°. However, if a sample is drawn from beyond the plateau, the contact angle value would result in greater than 90°.

In all, from the results analyzed above, it was observed that calcium carbonate, in water, is positively charged and has the potential to adsorb particles onto its surface. Oil, on the other hand, consists of negatively charged components which adsorb onto the calcium carbonate surface due to electrostatic interaction.

In a strongly water-wetted calcium carbonate system, upon the addition of crude oil, the zeta potential is observed to decrease until oil particles adsorb onto the rock surface. Effective inhibitors in such systems have been able to increase the surface resistance to adsorb negatively charged oil particles. In a strongly oil-wetted system, the zeta potential fluctuates upon the addition of water and eventually leads to the formation of large chunks of calcium carbonate with asphaltene deposition. Inhibitors have been found to be ineffective in such systems.

As evident from the different wettability studies carried out, crude oil adsorbs onto the limestone surface, initially, with a steady decrease, followed by a steep decrease in the zeta potential value and a plateau. On the other hand, maltenes have been observed to adsorb onto the surface with a steady decrease in the zeta potential value, while asphaltene molecules have been observed to readily adsorb onto the limestone surface, resulting in a steep decrease in the zeta potential value, followed by a plateau. The superimposition of zeta potential values for the wettability profiles shows a strong relationship that will be explained, in detail, in the following section.

3.3 Wettability alteration due to asphaltene

All wettability studies, carried out, indicate the adsorption of oil particles when crude oil or its individual components are added to a limestone-water mixture. However, a detailed explanation is required for such a behavior and the role of asphaltene in the wettability alteration process. **Figure 12** provides a mechanism that the oil components undergo through the wettability alteration process, at a microscopic level.

Initially, the limestone surface in the limestone-water system is completely water-wet. When crude oil is added, the resins, major oil component in the light fraction of oil (i.e., maltene), competitively adsorb onto the limestone surface. Resins are negatively charged in oil with a dipole moment of 2.4–3.2 D, while water has an average dipole moment of 1.85 D [32, 35]. Due to greater forces of electrostatic attraction between the positive limestone surface and negatively charged resins, the resins adsorb onto the limestone surface. At the end of the steady decrease, as indicated in **Figure 12**, a monolayer of resins is formed, pushing the water molecules to the bulk.

Following the steady decrease of zeta potential, the limestone surface experiences a steep decrease in zeta potential value. This is attributed to the competitive adsorption between the resins and the asphaltene molecules. Asphaltene molecules, as reported earlier, have a dipole moment over the range of 3.3–6.9 D [36]. This signifies the ability of asphaltene molecules to carry a greater charge than resins. Therefore, during the wettability alteration process, asphaltenes carry a greater negative charge than resins and competitively adsorb onto the limestone surface,

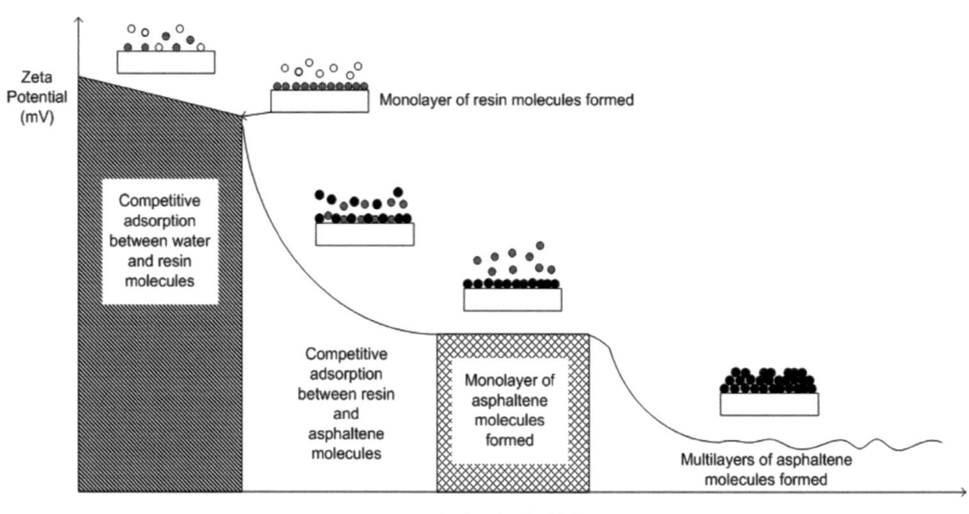

Figure 12.
Mechanism oil particles undergo during a wettability alteration process.

displacing the resin molecules to the bulk. The wettability curve flattens out when the asphaltene molecules form a monolayer on the limestone surface. Wettability studies of limestone for crude oils, A and B, tend to follow this behavior. Moreover, wettability of limestone for all asphaltenic solutions except 5 wt% is believed to follow this mechanism. According to a study conducted by Plank and Bassioni [37], $CaCO_3$ was found to correspond to a Type II isotherm [38].

According to Gregg et al. [39], the Type II isotherm, observed in physical adsorption, indicates the formation of a monolayer of particles onto the surface near the inflection point in the isotherm. Moreover, it allows for multiple layers to be formed above that monolayer of particles. The additional phenomenon, indicated in **Figure 12**, is believed to be observed in the wettability curve of limestone for crude oil C and is the reason behind the difference between the wettability of limestone for all asphaltenic solutions and 5 wt%. As the concentration of crude oil increases in the limestone-water crude oil system, another steep decrease in the zeta potential value is observed. The curve flattens out, and fluctuations are observed toward the end of the wettability curve. Discussing the mechanism following the monolayer formation of asphaltene molecules on limestone, it is believed that more asphaltene molecules approach the monolayer and adsorb onto it. This results in the formation of multilayers, leading to asphaltene deposition. This behavior can be attributed to the adsorption isotherm IV [38].

As described by Gregg et al. [39], the points of inflection in the adsorption isotherm type IV reflect the completion of monolayer as well as the onset of multilayer adsorption. Moreover, capillary condensation is associated with this isotherm where the adsorbate fills the small pores of the solid. In the case of the wettability alteration by crude oil, the pores can be assumed to be filled by asphaltenes which causes fluctuations. The wettability study, of limestone-oil system with water added, shows fluctuations at a significant level where oil is destabilized by water the system. Significant fluctuations, toward the end of the wettability profiles, are also evident in **Figure 4**, the wettability study of limestone for crude oil C, and evident in **Figure 10**, the wettability study of limestone for 5 wt% asphaltenic solution. When a concentrated liquid coats a porous solid, it tends to permeate through the solid rock and fill its pores. As observed from the wettability study of limestone-oil system with water added, water presence in such a system causes deposition of heavy fractions, leading to instability in the system.

Commonly, physical adsorption gives rise to such an isotherm resulting in multilayer adsorption. Components, comprising of aromatic rings, in the crude oil adsorb onto the monolayer of oil particles adsorbed on the $CaCO_3$ surface. As multilayers adsorb, the system experiences a steady decrease in the zeta potential value until a plateau is reached. Due to the multilayer adsorption, the thickness is believed to increase causing the CVI to measure a potential value at a distance beyond the electric double layer.

4. Conclusion

Complete wettability profiles of charged surfaces can be generated using zeta potential measurements. Unlike the commonly employed methods, the novel technique is able to study wettability in situ alteration. In the case of calcium carbonate and crude oil, the tolerance of the reservoir rock surface was observed while transitioning from a water-wet to an oil-wet surface. In this chapter, a comprehensive insight for the mechanism of inhibition of asphaltenes by inhibitors was provided. Moreover, a detailed discussion is provided on the role of asphaltene in wettability alteration, by stating the mechanism it follows in order to form multilayer adsorption.

According to literature, factors such as temperature, pressure, and pH have an effect on reservoir rock wettability. Therefore, further research can be carried out in order to study the effect of these factors on the wettability alteration process. Moreover, as the wettability study using zeta potential measurements is a novel technique, various other studies, in which traditional wettability methods were used, can be carried out with this approach. Additionally, a similar analysis can be carried out with different types of reservoir rocks, and wettability profiles can be generated.

Acknowledgements

Figures 3, **5**, **11**, and **12** have been adapted with permission from Ref. [33]. Copyright 2016 American Chemical Society.

Author details

Syed Taqvi[1] and Ghada Bassioni[2]*

1 University of Waterloo, Waterloo, Canada

2 Ain Shams University, Cairo, Egypt

*Address all correspondence to: ghada_bassioni@eng.asu.edu.eg

References

[1] Bortolotti V, Macini P, Srisuriyachai F. Wettability index of carbonatic reservoirs and EOR: Laboratory study to optimize alkali and surfactant flooding. In: SPE International Oil and Gas Conference and Exhibition in China; 2010. p. 12

[2] Donaldson EC, Thomas RD, Lorenz PB. Wettability determination and its effect on recovery efficiency. Society of Petroleum Engineers Journal. 1969;**9**(1):13-20

[3] Agbalaka CC, Dandekar AY, Patil SL, Khataniar S, Hemsath J. The effect of wettability on oil recovery: A review. In: SPE Asia Pacific Oil and Gas Conference and Exhibition; SPE; 2008. pp. 1-13

[4] Glover P. Chapter 07: Wettability, Formation Evaluation M.Sc. Course Notes, Vol. 7, pp. 76-83; 2001

[5] Fatt I, Klikoff WA. Effect of fractional wettability on multiphase flow through porous media. Journal of Petroleum Technology. 1959;**11**(10):71-76

[6] Mohanty KK, Salter SJ. Multiphase flow in porous media: III oil mobilization, transverse dispersion and wettability. In: SPE 58th Annual Technical Conference and Exhibition; 1983. pp. 1-21

[7] Lyons WC, Plisga GS, Lorenz MD. Reservoir engineering. In: Lyons WC, Plisga GS, Lorenz MD, editors. Standard Handbook of Petroleum and Natural Gas Engineering. 3rd ed. Elsevier; 2016. pp. 5-1-5-291

[8] Zinszner B, Pellerin FM. A Geoscientist's Guide to Petrophysics, TECHNIP. Paris: IFP Publications; 2007

[9] Taqvi ST, Almansoori A, Bassioni G. Modeling the impact of wettability alterations on calcium carbonate system for crude oil and asphaltenic solutions. Industrial and Engineering Chemistry Research. 2014;**53**(12):4773-4777

[10] Hodne H. Rheological Performance of Cementitious Materials Used in Well Cementing. [Doctoral thesis]. Stavanger: University of Stavanger; 2007

[11] Han KN. Fundamentals of Aqueous Metallurgy. Littleton: Society for Mining, Metallurgy, and Exploration, Inc; 2002

[12] BTG Instruments GMBH. Product Sheet Mütek PCD-04 Particle Charge Detector. Wessling; 2007

[13] Talero R, Pedrajas C, Rahhal V. Performance of fresh Portland cement pastes—Determination of some specific rheological parameters. In: Durairaj R, editor. Rheology: New Concepts, Applications and Methods. 1st ed. London: IntechOpen; 2013. pp. 57-79

[14] Malvern Instruments. Zetasizer Nano Series User Manual. Worcestershire: Malvern Instruments Ltd.; 2004

[15] Salopek B, Krasić D, Filipović S. Measurement and application of zeta-potential. Rudarsko-Geološko-Naftni-Zbornik. 1992;**4**:147-151

[16] Dukhin AS, Ohshima H, Shilov VN, Goetz PJ. Electroacoustics for concentrated dispersions. Langmuir. 1999;**15**(10):3445-3451

[17] Moulin P, Roques H. Zeta potential measurement of calcium carbonate. Journal of Colloid and Interface Science. 2003;**261**(1):115-126

[18] Patil S, Sandberg A, Heckert E, Self W, Seal S. Protein adsorption and cellular uptake of cerium oxide nanoparticles as a function

of zeta potential. Biomaterials. 2007;**28**(31):4600-4607

[19] Schellman JA, Stigter D. Electrical double layer, zeta potential, and electrophoretic charge of double-stranded DNA. Biopolymers. 1977;**16**(7):1415-1434

[20] Bangham AD. The adhesiveness of leukocytes with special reference to zeta potential. Annals of the New York Academy of Sciences. 1964;**116**:945-949

[21] Zhang Y et al. Zeta potential: A surface electrical characteristic to probe the interaction of nanoparticles with normal and cancer human breast epithelial cells. Biomedical Microdevices. 2008;**10**(2):321-328

[22] Weidenhammer P, Jacobasch HJ. Investigation of adhesion properties of polymer materials by atomic force microscopy and zeta potential measurements. Journal of Colloid and Interface Science. 1996;**180**(1):232-236

[23] Bassioni G. A study towards 'greener' construction. Applied Energy. 2012;**93**:132-137

[24] Bassioni G, Ali MM. Studying the physico-chemical properties of commercially available oil-well cement additives using calorimetry. Journal of Thermal Analysis and Calorimetry. 2013;**111**(1):295-303

[25] Bassioni G. Mechanistic aspects on the influence of inorganic anion adsorption on oilfield scale inhibition by citrate. Journal of Petroleum Science and Engineering. 2010;**70**(3-4):298-301

[26] Bassioni G, Taha Taqvi S. Wettability studies using zeta potential measurements. Journal of Chemistry. 2015;**2015**:1-6

[27] Elimelech M, Chen WH, Waypa JJ. Measuring the zeta (electrokinetic) potential of reverse osmosis membranes by a streaming potential analyzer. Desalination. 1994;**95**(3):269-286

[28] Bassioni G, Ali MM, Almansoori A, Raudaschl-Sieber G, Kühn FE. Rapid determination of complex oil well cement properties using mathematical models. RSC Advances. 2017;**7**(9):5148-5157

[29] Bousse L, Mostarshed S, Van Der Shoot B, de Rooij NF, Gimmel P, Göpel W. Zeta potential measurements of Ta_2O_5 and SiO_2 thin films. Journal of Colloid and Interface Science. 1991;**147**(1):22-32

[30] Jayaweera P, Hettiarachchi S, Ocken H. Determination of the high temperature zeta potential and pH of zero charge of some transition metal oxides. Colloids and Surfaces A: Physicochemical and Engineering Aspects. 1994;**85**(1):19-27

[31] Anderson WG. Wettability literature survey-part 3: The effects of wettability on the electrical properties of porous media. Journal of Petroleum Technology. 1986;**38**(12):1371-1378

[32] Yan J, Plancher H, Morrow NR. Wettability changes induced by adsorption of asphaltenes. SPE Production & Facilities. 1997;**12**(4):259-266

[33] Taqvi ST, Almansoori A, Bassioni G. Understanding the role of asphaltene in wettability alteration using ζ potential measurements. Energy & Fuels. 2016;**30**(3):1927-1932. DOI: 10.1021/acs.energyfuels.5b02127

[34] Gonzalez G. Surface charge and potential at the asphaltenes-solution interface. In: 8th International Conference on Petroleum Phase Behavior and Fouling; 2007

[35] Clough SA, Beers Y, Klein GP, Rothman LS. Dipole moment of water from Stark measurements of H_2O, HDO,

and D_2O. The Journal of Chemical Physics. 1973;**59**(5):2254-2259

[36] Goual L, Firoozabadi A. Effect of resins and DBSA on asphaltene precipitation from petroleum fluids. AICHE Journal. 2004;**50**(2):470-479

[37] Plank J, Bassioni G. Adsorption of carboxylate anions on a $CaCO_3$ surface. Zeitschrift für Naturforschung - Section B Journal of Chemical Sciences. 2007;**62**(10):1277-1284

[38] Lowell S, Shields JE, Morral JE. Powder surface area and porosity, 2nd edition. Journal of Engineering Materials and Technology. 1985;**107**(2):180

[39] Gregg SJ, Sing KSW, Salzberg HW. Adsorption surface area and porosity. Journal of the Electrochemical Society. 1967;**114**(11):279C

Chapter 4

Wetting of Al Alloys for Hot Dipping Coating Process

Qiaoli Lin, Ran Sui and Weiyuan Yu

Abstract

Wetting phenomenon, as a basic physical process, also relates to the many material processes, such as coating process, brazing or soldering process, casting, preparation of MMCs, etc. In this chapter, the method of wetting characterization at the high temperatures was presented (especially for the metallurgical melts), also the wetting behavior and mechanism of Al alloys (4043 and 6061 alloys) on the different metallic substrates at isothermal dwelling process as well as the characteristics and formation mechanism of precursor film were discussed. Thermodynamics of segregation of solute element were also discussed, which can be predicted by the thermodynamic model. We believe the content of this chapter would be a guidance for hot dipping process based on the wetting theories.

Keywords: wettability, aluminizing, brazing, interfacial reaction, adsorption, precursor film

1. Introduction

At high temperatures, wetting of a solid (metal or ceramic) by the molten metallurgical melts is of great technological importance in a variety of metallurgical processes, e.g, hot dipping coating process, brazing process, casting process and sintering process. Each process has the different optimal wetting condition. For the hot dipping coating process, the perfect wettability of base materials by liquid metal is demanded. Two key issues for wetting at high temperatures include the spreading dynamics (wetting behavior) and the final wettability (the degree of wettability). For the former issue, it determined the technological parameters in process; for the latter issue, it would be one of the critical evaluation bases for whether the process can be carried out or not.

The description of wettability since from 1805 in Young's work [1], has been well developed, as following,

$$\cos\theta = \frac{\sigma_{sv} - \sigma_{sl}}{\sigma_{lv}} \quad (1)$$

where θ is contact angle for a liquid equilibrium with ideal solid at the point of triple line, σ_{lv}, σ_{sv} and σ_{sl} are the tension between liquid-vapor, solid-vapor and solid-liquid interfaces. The value of θ represents the degree of wettability, i.e., the better wettability and the smaller θ, also $\cos\theta$ stands for the energy in a dimensionless form. The methods for obtaining θ are numerous, but for the wetting system of metallurgical melts at high temperatures, need to consider many other factors, not just the high temperatures, metallurgical process, atmosphere, interfacial reaction, etc. Further, the characterization of wetting should pay attention to two points, one is the wetting under the quasi-ideal condition, another is the wetting in a real process which including the consideration of many other factors, such as nonequilibrium temperature field, heat input and output, physic-chemistry effect of flux on interface, etc. Due to the complexity, the method for the characterization of the wetting under the quasi-ideal condition would be elaborated in Section 2.

The coating process by using a hot dipping method, although a traditional technology with relatively high energy consumption, is also a reliable technology with high efficiency. The quality of coating directly depends on the wetting of base metal by the coating metal. The performance of coating as well as the technological parameters in the coating process is also affected by the trace addition in the alloys which may act on the solid/liquid interface and(or) the liquid/vapor interface. In this chapter, based on the effect of trace elements on the wetting behaviors and the interfacial structures, the wetting mechanism of base metals (steel, Ti6Al4V (TC4) alloy, pure Ti (TA2)) by Al alloys (4043 alloy and 6061 alloy) as well as the formation of precursor film in these systems would be presented in Section 3.

This chapter presents an overview of wetting parameters at high temperatures, wetting mechanism in the isothermal spreading, the effect of trace elements on the formation of interface and the possibility of designed interfacial structures.

2. Characterization of wetting behavior

At high temperatures, one needs to be careful in the characterization of the wetting of molten metal on the surface of materials (ceramics or metals) with relatively high melting pointing under the quasi-ideal condition.

First, the basic condition is the homogeneous temperature field for wetting should be established.

Second, the average roughness (R_a) of substrate should be obtained before wetting test. Usually, R_a for the metallic substrates after polishing and the monocrystal ceramic substrates are in the range of dozens of nanometers over a distance of 2 mm, R_a for the polycrystal ceramic substrates after polishing are in the range of hundreds of nanometers which depends on the relative density of the polycrystal ceramic substrates.

Third, due to the high sensitivity of metal to oxygen, the oxygen partial pressure in the atmosphere should be controlled, usually in a vacuum of ~10^{-4} Pa or a dried-deoxidized Ar atmosphere, the oxygen partial pressure was estimated to be lower than 10^{-14} Pa [2] in such a high vacuum, and the oxygen partial pressure was estimated to be lower than 10^{-8} Pa [2] in a dried-deoxidized Ar atmosphere.

Once the above conditions were confirmed, and then the various methods for the testing should be considered.

Several methods for the testing at room temperatures, such as sessile drop method, wetting balance method, vertical rod method, tilted plate method, capillary rise method, etc., but the sessile drop method based on the complicated calculation is usually adopted due to the good feasibility. One also should note that the sessile drop method used in more than 90% of wetting studies at high temperatures [3]. The size of drop for testing should refer to its capillary length (i.e., $2\sigma_{lv}/(\rho g)^{1/2}$, where ρ is the density of liquid and g is the acceleration due to gravity), and thus

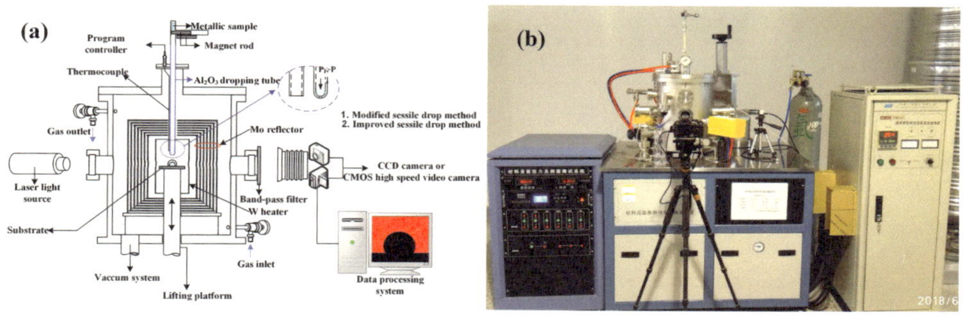

Figure 1.
(a) Schematic diagram for wetting furnace and (b) the designed wetting furnace.

a drop of volume ~5 μl with a spherical diameter of ~2 mm will be suitable for obtaining of the contact angle. Advancing contact angle is the parameter which the engineers mostly care about, and thus the study of wetting behavior used the advancing contact angle.

To acquire wetting parameters under the quasi-ideal condition, a specific wetting furnace is necessary, as be shown in **Figure 1**. The whole device mainly includes the controlled heat system, the vacuum system, the inertia gas flow system, the water-cooling system, and the image acquisition and data processing system. The details of this device were shown in **Figure 1(a)**, W, Ta or Mo was used as a heater, and two typical drop transfer modes were used, the modified sessile drop method and the improved sessile drop method.

Comparing to the traditional sessile drop method (substrate and metal were heated together), the mentioned two sessile drop methods have a distinct advantage in measurement of initial contact angle and spreading dynamics, particularly for the system with a chemical reaction at elevated temperatures.

To acquire the wetting parameters accurately, a laser backlight source (650 nm in wavelength and 10 mW in power) together with the band-pass filter is necessary for an image acquisition and data processing system. To insure the reliability of obtained data, the drop profiles were captured in two axes, as shown in **Figure 1(b)**. The typical side view of captured drop profile was shown in **Figure 2**. Before the calculation of contact angle, every pixel in the image was defined as a coordinate value (x and y coordinates), and then the program can seek the boundary between drop and backlight, and thus the pixel coordinate function of drop profile can be

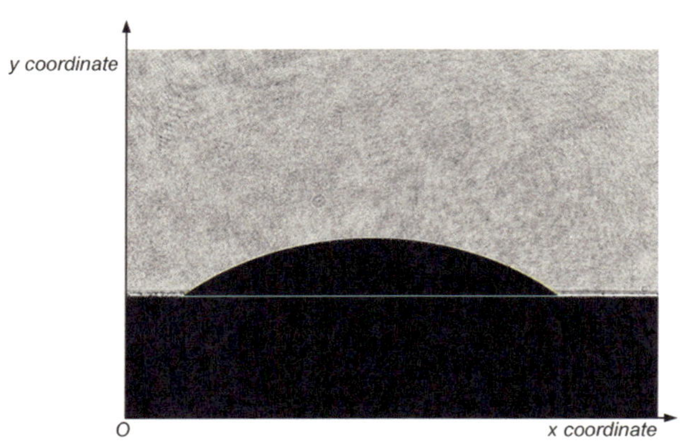

Figure 2.
Typical captured drop profile.

	Si	Fe	Cu	Mn	Mg	Zn	Ti	Al
Al 6061	0.60	0.90	—	—	0.90	0.25	0.15	Bal.
Al 4043	4.5–6.0	0.80	0.30	0.05	0.50	0.10	0.20	Bal.

Table 1.
Nominal chemical compositions of the used Al alloys (wt.%).

Steel	C	Si	Mn	P	S	Fe	
Q235	0.12–0.2	≤0.3	0.3–0.7	≤0.045	≤0.045	Bal.	
Ti alloy	Ti	Fe	Si	C	Al	V	O
TC4	Bal.	0.3	—	0.1	5.6–5.8	3.5–4.5	0.2
Pure Ti	Ti	Fe	Si	C	N	H	O
TA2	Bal.	0.3	0.15	0.1	0.05	0.015	0.2

Table 2.
Nominal chemical compositions of the used substrates (wt.%).

extracted. The function was further fitted by several models, i.e., Young-Laplace model, conic model, circle model. The choice of fitting model depends on the symmetry of captured drop profile or the contact angle. If the drop profile is almost axisymmetric, Young-Laplace model is the first choice. If the profile deviates axis symmetry, conic model is the first choice. If the contact angle is less than 5°, conic model or circle model is the first choice. Based on the results of calculation, the detailed wetting parameters can be obtained, i.e., the contact angle, the base diameter, the density, volume and surface tension of liquid.

The chemical compositions of the used materials were shown in **Tables 1** and **2**.

Based on the wetting characterization, to establish the relationship of wetting behavior and surface/interface evolution, further reveal the wetting mechanism, some necessary micro-analysis were also carried out.

3. Reactive wetting of Al alloys on metallic substrates

Trace elements in Al alloys, which may act on the surface of drop and/or the solid/liquid interface, and thus the industrial grade Al 4043 alloy and Al 6061 alloy were selected in this work which are of Si addition (~5 wt.%) and of Mg addition (~1 wt.%), respectively. The Si addition may both act on the surface of drop and solid/liquid interface, and the Mg addition may act on the surface of the drop mainly. The metallic substrates were selected as a reactive wetting system, i.e., Al/steel (low-carbon steel), Al/TC4 (Ti6Al4V) and Al/TA2 (TA2 grade pure Ti) systems.

The melting points of Al 4043 alloy and Al 6061 alloy were confirmed by DTA method (STA449-C, NETZSCH, and Germany), are 586 and 582°C, respectively, and thus the isothermal wetting experiments were carried out in a high vacuum (~10^{-4} Pa) by using improved sessile drop method at the range from 600 to 700°C.

The variations in the contact angle and the normalized contact radius (R_d/R_0, R_d is the dynamic contact radius and R_0 is the initial contact radius) with time are critical parameters for fabricating a coating by hot dipping method, were shown in **Figures 3** and **4**. **Figure 3** for Al/steel, **Figure 4(a)–(d)** for Al/TC4, **Figure 4(e)–(h)** for Al/TA2, all show the monotonic variation with time and the temperature depended final wettability. In variations of normalized contact radius with time, there are two types of variations, i.e., the linear spreading and

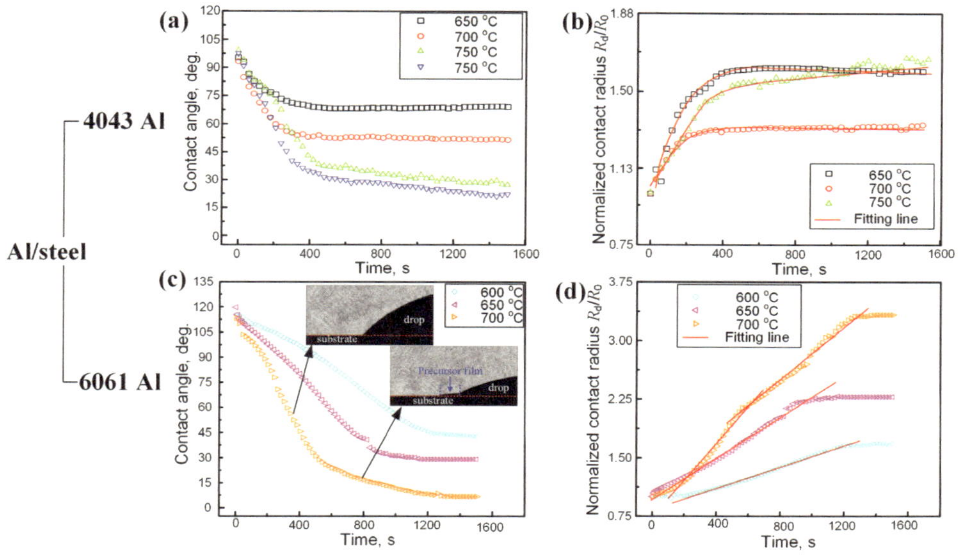

Figure 3.
Variation in contact angle and normalized contact radius with time, (a) and (b) for 4043 Al/steel, (c) and (d) for 6061 Al/TC4, respectively.

the nonlinear spreading. In Al/steel system, the variation of Al 4043 is in the nonlinear spreading, and the variation of Al 6061 is in the linear spreading; In Al/TC4 and Al/Ti systems, the variations of R_d/R_0 show that nonlinear spreading firstly, and then linear spreading. The variations of the nonlinear spreading can be described by the exponential function ($R_d/R_0 = R_f/R_0\text{-}a\exp[-(t/\tau)^m]$, where a, τ and m are the fitting parameters, R_f represents the final contact radius).

During spreading, the precursor film (diffusion bond or so called "wetting halo") was formed at the latter stage of spreading in some specific samples. As shown in **Figure 5(a)–(c)** are the top-views for the samples after isothermal wetting at 650, 700 and 750°C for Al 4043/steel, and the top-views of (a-1) to (c-1) for Al 6061/steel samples after isothermal wetting at 600, 650 and 700°C. As a consensus, the formation of the precursor film always accompanies the good final wettability, and the larger width of the precursor and the better final wettability. Therefore, the final wettability of Al 6061/steel is better than Al 4043/steel at the same experimental temperature, although the latter has some precursor film at 650 and 700°C, the width of them is very limited (~500–1000 μm). Such a congruent relationship is also suitable for the wetting of metal/ceramic systems [4]. Also, in all Al/TC4 and Al/TA2 (except for Al 6061/TA2 at 600°C which is nonwetting at the final state), the precursor film can be found at the final state after wetting, also can be seen in Hg/Ag system [5] in the study by Be'er et al. However, the formation mechanism is so different from the evaporation-condensation mechanism (suggested by de Gennes [6]) and the surface diffusion mechanism (suggested by Li et al. [7]). Further, the precursor film, in a metal/ceramic system, especially for the melt contain some concentration of the active element, is an adsorbed film. The formation of the film should satisfy some specific conditions, as we reviewed before [4].

The sectional views of interfacial structures for Al 6061/steel and Al 4043/steel samples are obviously different after isothermal wetting at the same temperature (at 650°C), as shown in **Figures 6(a)** and **7(a)**. Although two types of Fe-Al intermetallics can be found in all the samples (bottom: the continuous Fe_2Al_5 layer, upper: discontinued $FeAl_3$), the effect of trace elements on the interfacial structures is obvious. In Al 4043/steel, the addition of Si segregated at solid/liquid interface

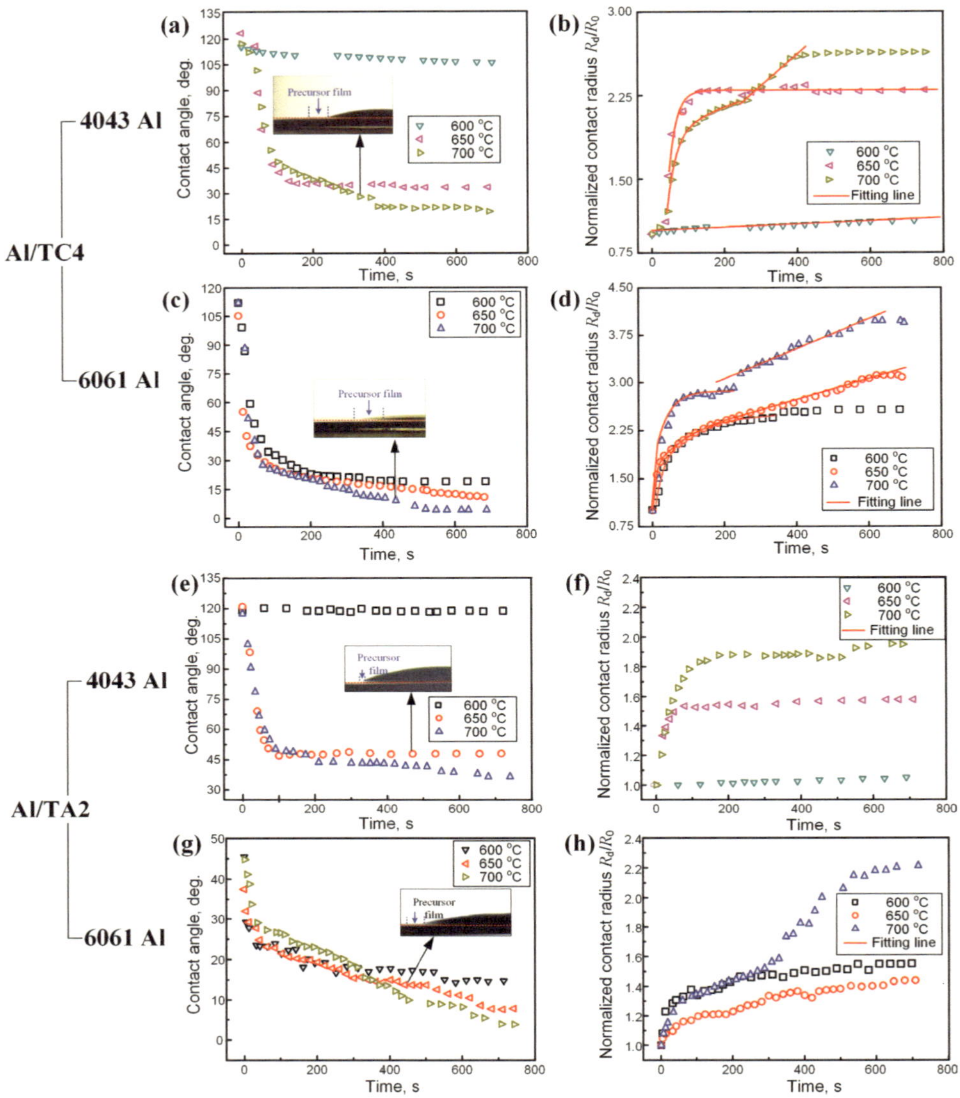

Figure 4.
Variation in contact angle and normalized contact radius with time, (a) and (b) for 4043 Al/TC4, (c) and (d) for 6061 Al/TC4, (e) and (f) for 4043 Al/TA2, (g) and (h) for 6061 Al/TA2, respectively.

which enhanced the interfacial reaction, so that the barrier of interdiffusion was established. Both Si element distribution map (**Figure 6(c)**) and the elements line distributions for the corresponding position of **Figure 6(b)** show Si is incline to segregate at solid/liquid interface rather than surface of liquid. Also, the segregation of Si induced the brittleness of the compact Fe_2Al_5 layer [8], which induced some continuous and propagating cracks. Actually, some Si was dissolved into the Fe_2Al_5 phase as a solid solution due to the Si segregation, which can be rewritten as $Fe_2(Al_{1-x}Si_x)_5$, where x is in the range of 0.0625–0.104, as reported by Gupta [9]. Comparing with Al 4043/steel, in the interface of Al 6061/steel, no Mg segregation (even no trace of Mg) can be found in the interface or the bulk of drop due to the high volatility of Mg under the vacuum condition, as shown in **Figure 7(a)–(c)**. The Fe_2Al_5 reaction layer on steel side is irregular and stretches into the steel side, which is one of the main differences from Al 4043/steel interface. Without the barrier

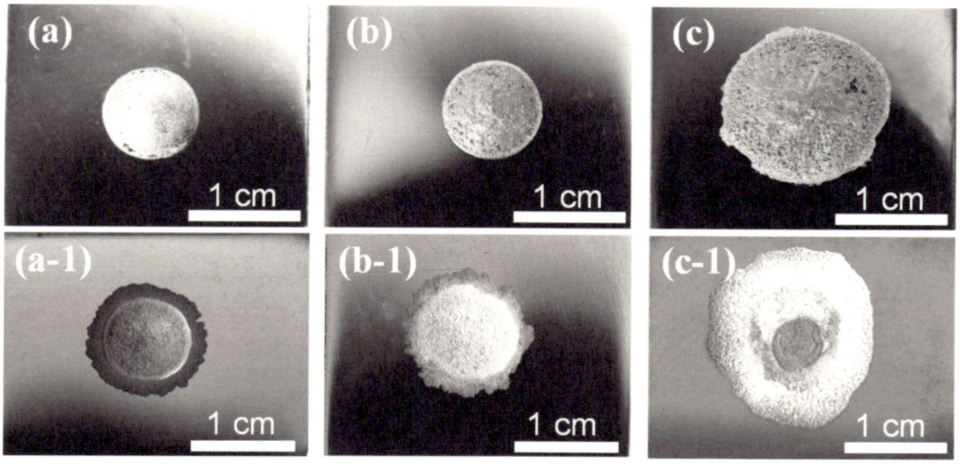

Figure 5.
Top-view of some typical sessile drop samples of Al/steel after the wetting experiments in vacuum: (a)–(c) for Al 4043 sample at 650, 700 and 750°C; (a-1)–(c-1) for Al 6061 sample at 600, 650 and 700°C.

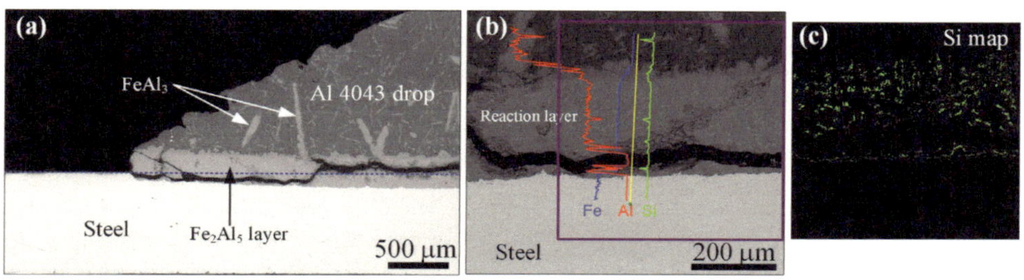

Figure 6.
Cross-sectional views of interface structures (Al 4043/steel): (a) at the close of triple line, (b) the central position at interface, (c) Si element distribution map corresponding to the purple rectangle in (b).

of interdiffusion, the grain boundary as a short-circuit diffused path, Al element prefers to diffuse into the grain boundary, and then reacted with Fe. The X-Ray micro-Diffraction (micro-XRD) pattern of the phases at the surface of the precursor film for Al 6061 sample after isothermal wetting at 700°C in **Figure 7(d)** shows the precursor film contains Fe_2Al_5, i.e., the extended reaction layer.

Understanding the reasons caused the wetting, is the benefit of further controlling the manufactured process, which is the prime concern of engineering. In a metallic system, the first obstacle for wetting is the oxide film on the surface of substrate. Protsenko et al. [10] considered the effect of the formation of intermetallic compounds in the wetting with the metallic substrate covered by oxide film. By the formation of IMC by diffusion of reacting components through the thin oxide layer, the oxide film can be disrupted and then in situ a clean surface of intermetallic for wetting can be created. Also, as suggested by Durandet et al. [11], the precipitation of Fe-Al intermetallics is so fast even the contact time of liquid Al and steel as short as 20 ms. Further, the thickness of the IMC layer in this work is larger than the oxide film (in nanoscale), and thus the precipitation of IMC may be a factor for improving wettability. However, the time for spreading cannot correspond to the fast reaction.

The dynamic of IMC precipitation should not be the limited factor for spreading. Although the nonwettable Fe—O oxide film would block wetting, a reduction reaction $2yAl + 3Fe_xO_y = yAl_2O_3 + 3xFe$ can take place based on the thermodynamic consideration [12], and then Al melts can further react with the fresh surface of Fe.

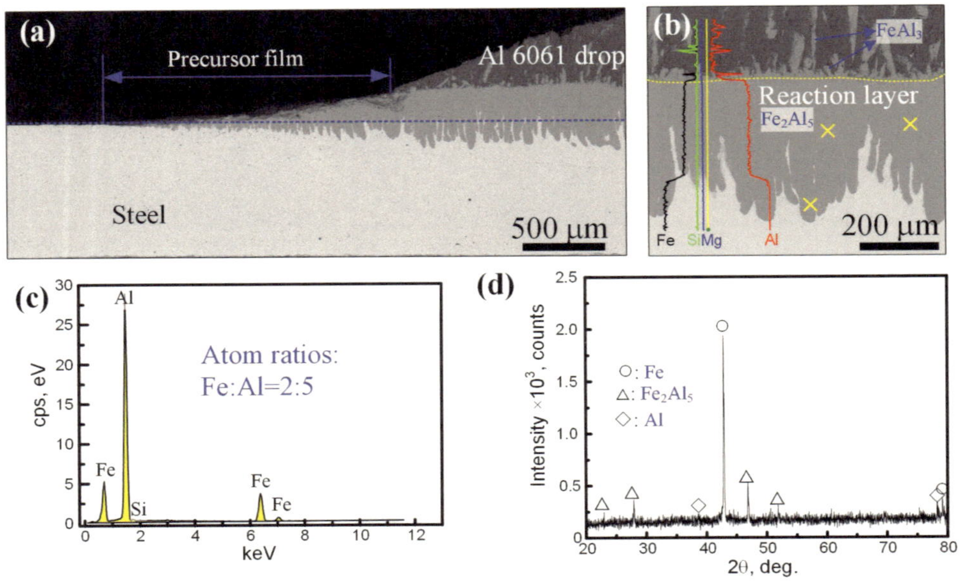

Figure 7.
Cross-sectional views of interface structures (Al 6061/steel): (a) at the close of triple line, (b) the central position at interface, (c) the typical energy spectrum for the position of yellow cross in (b) and (d) the XRD pattern for the surface of precursor film.

As known, the whole process is limited by the slowest step, and thus the reduction step may be the limited factor for spreading. In Al 6061/steel, although no trace of Mg was observed, the volatilization of Mg cannot be neglected. As suggested by Miller and Pa [13], Mg vapor as a gas flux can reduce Al_2O_3 in the brazing process of Al alloys. Here, the Fe—O oxide film also can be reduced by Mg vapor due to the more positive formation of Gibbs free energy of Fe—O compounds [12] under the same thermodynamic condition. Based on the result of microstructures, although the precursor film for Al 4043/steel also can be found, it only appears at high experimental temperature with very limited width due to the slight volatility of Al. All the interfacial structures indicate that the formation of the precursor film is related to the volatility of the active element in the specific system. However, the formation mechanism is not the evaporation-condensation mechanism. When the base metal was covered by the thin oxide film, the molten metal would infiltrate under the covered oxide film and then trigger the moving of triple line, and thus the spreading in this stage is also called secondary spreading or wetting, i.e., the formation of precursor film satisfies the subcutaneous infiltration mechanism (as proposed by Zhuang and Lugscheider [14]).

Based on the final wetting states, i.e., nonwetting and wetting states, the typical interfacial structures were selected, as shown in **Figures 8–10**. In **Figure 8(a)** and **(b)**, the precursor with width of several millimeters as well as some Al-Ti intermetallics in the precursor film can be found. The obvious precursor film in Al/TC4 and Al/TA2 (except for the Al/TC4 and Al/TA2 at 600°C which are nonwetting) has the same formation mechanism i.e., so-called "subcutaneous infiltration". The final wettability was less affected by the oxide film but was determined by the reaction products at the liquid/solid interface. As known, although some residual oxygen in the vacuum chamber could dissolve into Ti due to the high affinity of Ti to oxygen, and then inducing oxidation, however, the diffusion rate for oxygen into the interior of Ti is faster than the oxidation rate so that the oxide film would be thinned or even removed. In Al/TC4 and Al/TA2 systems, both the temperatures and the

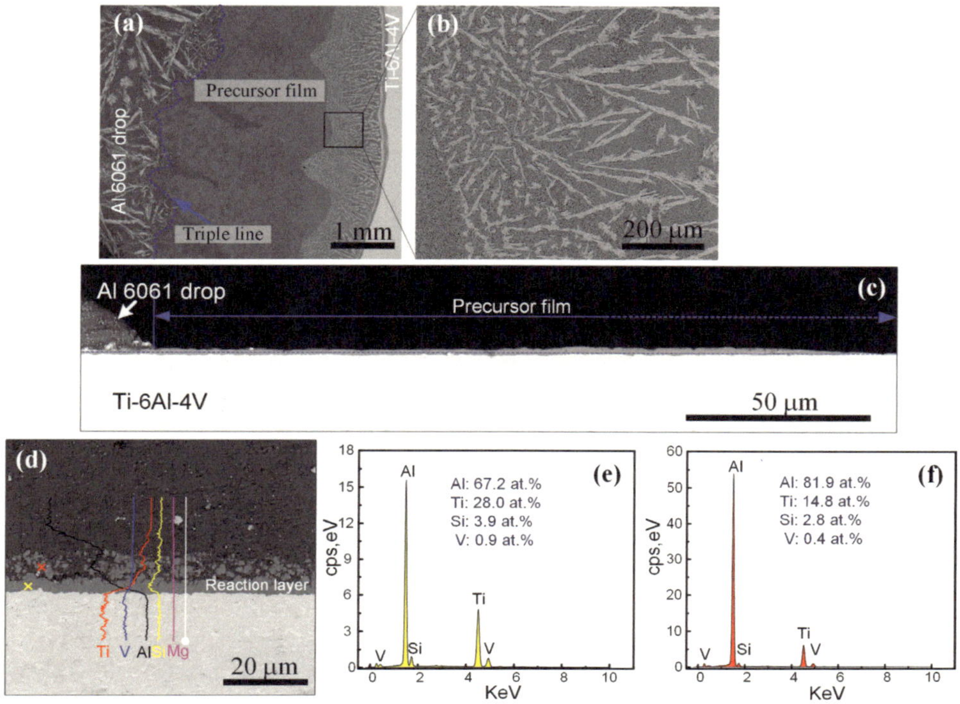

Figure 8.
(a) Top-view of the microstructures around the triple line for Al 6061/TC4 after isothermal wetting at 600°C; (b) detail of the black rectangle area in (a); cross-sectional view of Al 6061/TC4 (c) near the triple line and (d) at central position of the interfacial microstructures with the elemental line distribution results; (e) and (f) EDS results for the corresponding colored crosses in (d), respectively.

alloying elements (Si and Mg) caused the different interfacial structures and are responsible for the different final wettability. In **Figure 8(d)–(f)**, the typical cross-sectional view of Al 6061/TC4 after isothermal wetting at 600°C shows the chemical compositions of the granular phase distributed above the reaction layer are closed to Al_3Ti. The original concentration of Si in the bulk Al 6061 alloys is only 0.6 at.%, but the concentration of Si in the continuous reaction layer is far beyond that value, as shown in **Figure 8(e)**. The distinct segregation of Si in the reaction layer was formed during wetting. For the nonwetting sample of Al 4043/TC4 (after wetting at 600°C), as shown in **Figure 9(a)** and **(b)**, the interfacial structures with the virgate phase but no granular phase near triple line and the virgate phase together with granular phase at the center position can be found. Si concentration in the virgate phase is extremely high comparing with the original concentration in bulk Al 4043 alloy, as shown in **Figures 9(c)–(e)**.

For the wetting sample of Al 4043/TC4 which is similar to the interfacial structures in Al 6061/TC4, a precursor film with width of several millimeters was formed. However, such a precursor film contains two layers, as be shown in **Figure 10(a)** and **(b)**. From the cross-sectional view of the film (**Figure 10(c)**), the upper layer is the residual Al, which was attracted by the capillary force from the loose reaction layer. At the central position of the interface (**Figure 10(d)**), the loose continuous layer also can be found, but the virgate phase almost disappeared and only can be found sporadically above the loose layer. Based on the EDS results (**Figure 10(e)** and **(f)**), the loose layer may be a continuous Al_3Ti layer dissolved with Si as a solid solution. For the sample of Al 4043/TC4 after isothermal wetting at 650°C, the solidified Al was removed by NaOH aqueous solution (1 mol/L), and

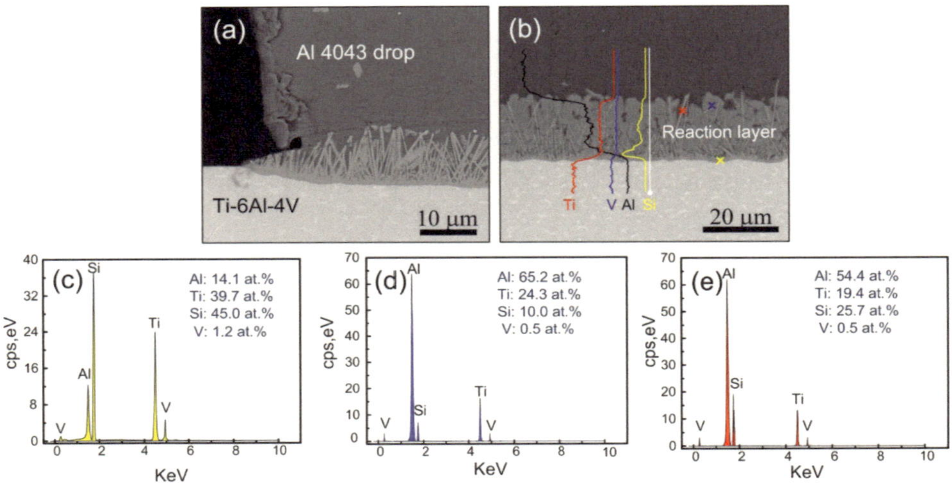

Figure 9.
Cross-sectional views for Al 4043/TC4 after isothermal wetting at 600°C: (a) at the triple line; (b) at the center of the interface; (c)–(e) the EDS results for the corresponding cross, circle and triangle in (b).

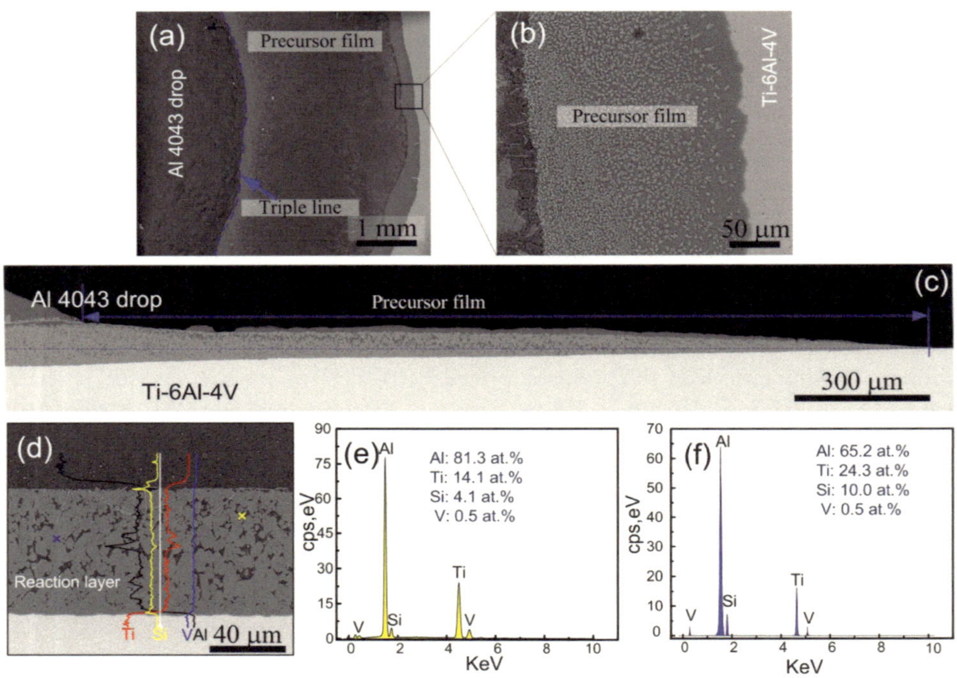

Figure 10.
(a) Top-view of the microstructures at the triple line for Al 4043/TC4 after isothermal wetting at 650°C; (b) detail of the black rectangle area in (a); cross-sectional view of Al 4043/TC4 (c) at the triple line, and (d) at center of the interfacial microstructures with the elemental line distribution results; (e) and (f) EDS results for the corresponding colored crosses in (d), respectively.

the macroscopical appearance of the sample and the details for the corresponding positions were shown in **Figure 11**. The granular phase in **Figure 11(b)** is corresponding to the loose continuous reaction layer, and the lamellate phase is corresponding to virgate phase in the cross-sectional view of the interface. Further, as shown in **Figure 12**, the XRD of the phases at the corresponding surface confirmed

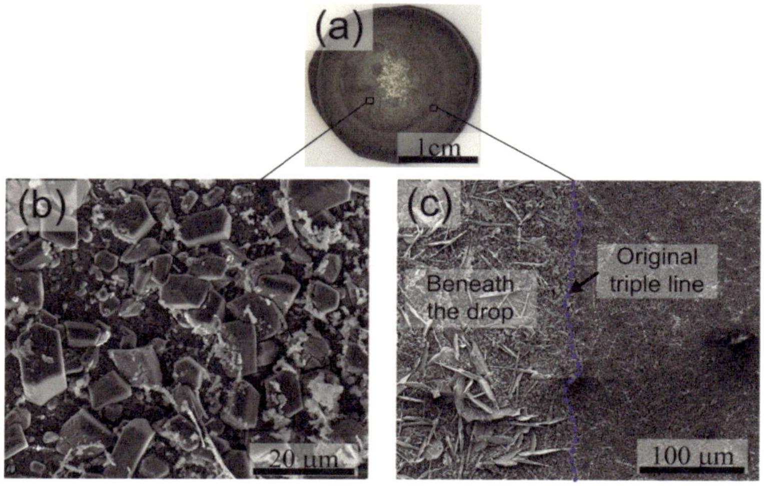

Figure 11.
(a) Solidified Al for the sample of Al 4043/TC4 after isothermal wetting at 650°C was removed by NaOH aqueous solution (1 mol/L); (b) and (c) the microstructures for the corresponding positions in (a).

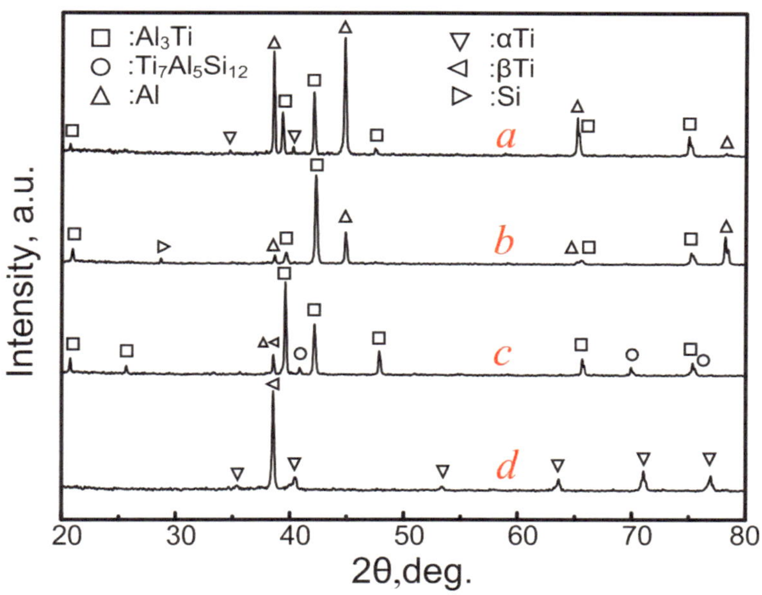

Figure 12.
XRD patterns of the phases at the precursor films: for (a) Al 6061 and (b) Al 4043 samples after isothermal wetting at 700°C; (c) the exposed interface of Al 4043/TC4 after isothermal wetting at 650°C through removing of the solidified Al drop using NaOH aqueous solution, and (d) the original surface of TC4.

the granular phase (the loose continuous reaction layer) is Al_3Ti, and the lamellate phase is $Ti_7Al_5Si_{12}$ (τ_1, a solid solution of Al in the $TiSi_2$ phase [15]).

The formed $Ti_7Al_5Si_{12}$ is a metastable phase, once the temperature was above 579°C, a decomposed reaction of $Ti_7Al_5Si_{12}$ would take place, i.e., $L + \tau_1 \leftrightarrow$ (Si) + (Al), at 579°C [15], where L is the liquid phase in composition (Al: 87.835, Si: 12.1, Ti: 0.065 in at.%), (Al) is in (~98.5 at.% Al and ~1.5 at.% Si). Once the continuous compact $Ti_7Al_5Si_{12}$ layer was formed, the mass transfer at the triple line would be further slowed, and the wettability was improved to a very limited degree (i.e., at 600°C). At higher temperature, $Ti_7Al_5Si_{12}$ was decomposed, and then the formation of Al_3Ti

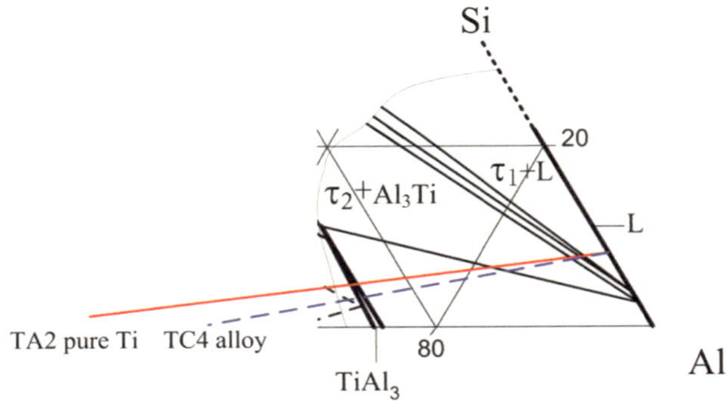

Figure 13.
Al-Ti-Si partial isothermal section at 700°C for Al rich corner [15].

replaced it. Hence, two layers (upper layer is some residual $Ti_7Al_5Si_{12}$, and under layer is Al_3Ti) in the precursor film of Al 6061/TC4 is not surprising, due to lack of L (i.e., the liquid phase) for the decomposed reaction. Also, because of the loosened structures of Al_3Ti, the reaction layer is no longer an obstruction (comparing to $Ti_7Al_5Si1_2$) and the mass transfer was intensified both at the interface and close of triple line, result in wettability was further improved. The interfacial structures of Al 6061/TA2 is similar to that of Al/TC4, so that we do not repeat them here. In this highly reactive wetting system, although the final wettability can be obtained, the interfacial reaction would never stop until the liquid phase was exhausted. Also, the interfacial structures of TA2 and TC4 seems no different, but Al 4043/TC4 alloy system is more incline to obtain Al_3Ti which is in a loosened structure (comparing to $Ti_7Al_5Si_{12}$ layer). As seen in the Al-Ti-Si ternary phase diagram (**Figure 13**) for Al rich corner at 700°C, the path from AlSi5 (i.e., Al 4043) to TC4 alloy (the blue dashed line) is shorter than that from AlSi5 to Ti (red line) as well as the distance of the crossed $\tau_1 + L$ region, which means Al 4043 (Al-5 wt.%Si) alloy contacted with TC4 alloy would induce the formation of Al_3Ti easily. Further, the loosened structure of Al_3Ti layer would cause the more mass transport and better wettability.

Si addition in Al alloys, as known, is a surface-active agent which can decrease the surface tension of the liquid and increase the flowability. However, in Al-Si/steel or Al-Si/Ti system, the interfacial microstructures confirm the affinity of Si to Fe (or Ti) is relatively higher than that of Si to the Al matrix. The Si segregation at the liquid/solid interface satisfies the thermodynamic condition. Such a thermodynamic model also can be used for predicting the segregation of alloying element at the liquid/solid interface. The adsorption energy based on the affinities, which can be described as following [3],

$$E_{M(B)}^{\infty SL} = m_1(\lambda_{AM} - \lambda_{BM} - \lambda_{AB}) \quad (2)$$

where $E_{M(B)}^{\infty SL}$ is adsorption energy, m_1 is an interfacial structure factor (always has a positive value), λ_{ij} is the mole exchange energy for i and j. Here, A, B and M stand for the solid metal, the liquid solvent metal and the solute metal, respectively. The more negative the value of $E_{M(B)}^{\infty SL}$, the solute element is more inclined to segregate at the liquid/solid interface. Further, the positive value of λ_{ij} means two metals is repellent in thermodynamic or weak interaction, or vice versa. We assumed that the binary solution is an infinite dilution regular solution, and λ_{ij} can be calculated from the following [3],

$$\lambda_{ij} = \left(\overline{\Delta H_{i(j)}}^{\infty} + \overline{\Delta H_{j(i)}}^{\infty}\right)/2 \quad (3)$$

where $\overline{\Delta H_{i(j)}}^{\infty}$ and $\overline{\Delta H_{j(i)}}^{\infty}$ represent the mixing enthalpy of i in j and j in i, based on Miedema's model, Ref. [16]. The chemical adsorption is usually considered as a prerequisite for the reaction (or for the precipitation of the reaction product) when the concentration of the adsorbate is below the saturation adsorption concentration. Therefore, $E_{M(B)}^{\infty SL}/m_1$ can be used for the prediction of interfacial structure before experiment.

The calculated $E_{M(B)}^{\infty SL}/m_1$ (−19.8 kJ/mol) has a negative value in Al-Si/Fe system which suggested that Si would segregate at liquid/solid interface and corresponding to some Si segregation in Fe_2Al_5 layer, but the value of $E_{M(B)}^{\infty SL}/m_1$ for the Al-Mg/Fe system is 158.9 kJ/mol. The calculated $E_{M(B)}^{\infty SL}/m_1$ (−73.4 kJ/mol) also has a negative value in the Al-Si/Ti system, but the value of $E_{M(B)}^{\infty SL}/m_1$ for the Al-Mg/Ti system is 221.4 kJ/mol. All the calculated results indicated the segregation at the liquid/solid interface satisfied the thermodynamic model. The E/m_1 for selected systems of the binary solution contacting with a metallic substrate were shown in **Table 3**, of which the segregation at interface or not were verified in our wetting experiments. The positive value of E/m_1 indicates the solute metal would not segregate at solid/liquid interface, and vice versa.

All the variation in spreading of Al/steel and Al/Ti indicated the characteristics of reactive wetting, i.e., the typical linear spreading as well as the interfacial reaction. Reaction-limited models of linear spreading and nonlinear spreading for the description of the wetting behavior, as proposed by Eustathopoulos group, were expressed as following,

$$\frac{dR_d}{dt} = k_1 = C\exp\left(\frac{\Delta E_a}{RT}\right) \text{ for linear spreading} \tag{4}$$

$$\cos\theta_e - \cos\theta_d = (\cos\theta_e - \cos\theta_0)\exp(-k_2 t) \text{ for nonlinear spreading} \tag{5}$$

where C is a constant which relates to the activity of a reactive solute, ΔE_a is the activation energy for spreading caused by a reaction which may partially contain the activation Gibbs energy of the reaction, R is the gas constant, and T is the temperature in K. The left term dR_d/dt, i.e., the spreading velocity, can be obtained by the linear fitting from **Figure 3(d)**, **(d)** and **(f)** (the slope after fitting, was denoted as k_1). where the θ_0, θ_d, and θ_e, respectively, represent the initial, dynamic, and equilibrium contact angles. k_2 is a kinetic constant which contains the activation energy and can be deduced from the fitting results based on the variations of $\theta - t$ in **Figures 3** and **4**. The logarithms, both of constants k_1 and k_2, were plotted as a function of $1/T$, i.e., the Arrhenius plots, were shown in **Figure 14**. The deduced activation energies from the slopes were shown in **Table 4**.

In Al/steel system, two types of spreading mode, i.e., the linear spreading in Al 4043/steel and nonlinear spreading in Al 6061/steel, can be found. The activation energy for Al 4043/steel can be deduced from the slopes, is 8 kJ/mol for Al 4043. Also, the activation energy for Al 6061 can be deduced from the fitting result of Eq. (5), is 86 kJ/mol. In the work of wetting and spreading of molten pure Al on the surface of mild steel, studied by Ishida [17], the linear kinetics were also observed, and the activation energy is 21.8 kJ/mol. Obviously, the trace elements in the systems influenced the spreading dynamics, and caused the different activation energies. In Al 4043/steel, the segregation of Si at liquid/solid interface enhanced the reactivity, and then might lead to relatively small activation energy. In Al 6061/steel, the liquid Mg in molten Al 6061 alloy does not react with Fe directly. The moving of the triple depends on the removing of oxide film on the surface of substrate. The reduction of Fe—O oxide film by the Mg in Al 6061 and the formation of Fe-Al intermetallics play

B-M/A	Mg-Al/Fe	Mg-Al/Ti	Mg-Zn/Fe	Mg-Si/Fe	Mg-Cu/Fe	Mg-Ni/Fe	Mg-Si/Ti	Mg-Al/Cu	Mg-Si/Cu	Al-Si/Fe
E/m_1 (kJ/mol)	−149	−212	−73	−133	6	−35	−249	−17	21	−20
B-M/A	Al-Zn/Fe	Al-Si/Ti	Al-Si/Cu	Cu-Ag/Si	Ni-Zn/Fe	Al-Cu/Fe	Cu-Si/Fe	Cu-Zn/Fe	Mg-Zn/Ti	Al-Mg/Ti
E/m_1 (kJ/mol)	43	−73	1.8	21	70	181	−118	−42	−137	221

Table 3.
E/m_1 *of the selected B-M/A systems calculated from Refs. [3, 16].*

DOI: http://dx.doi.org/10.5772/intechopen.81282

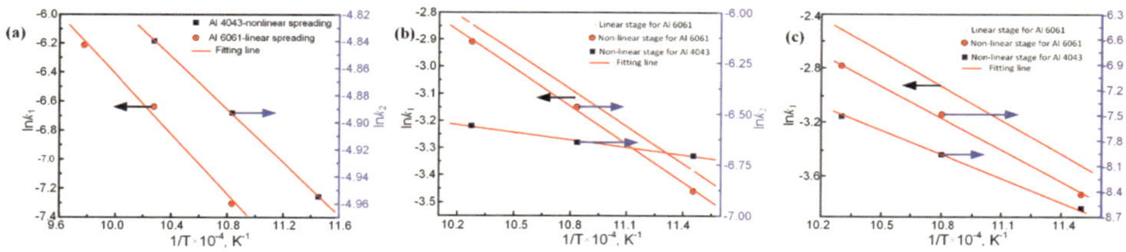

Figure 14.
Arrhenius plot of the kinetic constants k_1 and k_2: (a) Al/steel, (b) Al/TC4 and (c) Al/TA2.

Systems	Al6061/ steel	Al4043/ steel	Al6061/ TC4*	Al4043/ TC4	Al6061/ TA2*	Al4043/ TA2
E_a (kJ/ mol)	86	8	50	10	84	47

**The average activation energy of linear stage and nonlinear stage.*

Table 4.
Deduced activation energies from the slopes in **Figure 14.**

a combined action in the removing of oxide film, which can induce the spreading. Two spreading stages for Al 6061 at 700°C also indicate these two reaction mechanisms. Especially for the latter stage, the reduction reaction mechanism may play a major role on the moving of the triple line corresponding to the prolonged precursor film in this stage. Therefore, the apparent activation energy in Al 6061 is so different from Al 4043 and almost an order of magnitude larger.

In Al/TC4 system, such apparent activation energies should relate to the energy change of reaction. As suggested by Chen et al. [18], Gibbs energy changes of reaction per mole of reactants for the formation of $Ti_7Al_5Si_{12}$ and Al_3Ti in the range of 600–750°C are ~– 48 kJ/mol and ~– 30 kJ/mol, respectively. For the nonlinear stage, $Ti_7Al_5Si_{12}$ decomposition may play a dominant role in the change of interface tension corresponding to the apparent activation energy of 10–38 kJ/mol; for the linear stage, the decomposition of $Ti_7Al_5Si_{12}$ and the formation of Al_3Ti may play the role for the change of interface tension corresponding to the apparent activation energy of 62 kJ/mol. The relatively higher activity of Si in Al 4043 (due to the higher concentration) may slow down the decomposition of $Ti_7Al_5Si_{12}$, and then decrease the apparent activation energy. Further, the residual $Ti_7Al_5Si_{12}$ beneath the drop and above on Al_3Ti reaction layer at the close of the triple line (**Figure 11(c)**) should be one of the indicators that two reaction mechanisms together played a role on change of interface tension. Also, in Al/TA2 system, the apparent activation energies indicated that the whole wetting process is controlled by the above mechanism. Therefore, the activation energies (47–84 kJ/mol) may be also corresponding to the decomposition of $Ti_7Al_5Si_{12}$ and the formation of Al_3Ti.

4. Conclusions

Aluminizing of steel or Ti alloys can increase the ability of anti-corrosion, the service time and the working temperature, significantly. The wetting of Al alloys plays an important role in the aluminizing process, and thus studied in this work. The following conclusions can be drawn:

1. In Al/steel, the wettability was improved by intermetallic formation which would lead to the replacement of the oxidized surface by a clean surface of

an intermetallic compound, also due to the reduction by Al with oxide film; However, the final wettability of pure Ti (TA2) and Ti alloys (TC4) by Al was less affected by the oxide film, but was determined by the reaction products at the liquid/solid interface. Enhanced peritectic reaction of $Ti_7Al_5Si_{12}$ caused the different interfacial structures of TA2 and TC4.

2. The alloying elements in Al 6061 alloy (with Mg addition) and Al 4043 alloy (with Si addition) resulted in distinctly different interfacial structures, the formation of precursor film and spreading dynamics. Mg played a role like gas flux and reduced the oxide film on the surface of substrate. Si segregated at solid/liquid interface which satisfied the thermodynamic model. Such a thermodynamic model also can be used for predicting the element segregation at the interface.

3. The precursor film in these reactive wetting systems is an extended reaction layer. The formation of it satisfies subcutaneous infiltration mechanism.

4. The spreading dynamics of these systems can be described by RPC model, and the activation energies are related to the removing of oxide film covered the substrate or the reaction at interface.

Acknowledgements

This work is supported by National Natural Science Foundation of China (no. 51665031), "Kaiwu" Innovation Team Support Project of Lanzhou Institute of Technology (no. 2018KW-05).

Conflict of interest

We have no conflicts of interest to declare.

Author details

Qiaoli Lin[1*], Ran Sui[2] and Weiyuan Yu[1]

1 State Key Laboratory of Advanced Processing and Recycling of Non-ferrous Metal, Lanzhou University of Technology, Lanzhou, People's Republic of China

2 Department of Materials Science and Engineering, Lanzhou Institute of Technology, Lanzhou, People's Republic of China

*Address all correspondence to: lqllinqiaoli@163.com

References

[1] Young T. An essay on the cohesion of fluids. Philosophical Transactions of the Royal Society of London. 1805;**95**:65-87. DOI: 10.1098/rstl.1805.0005

[2] Yong-Taeg O, Fujino S, Morinaga K. Fabrication of transparent silica glass by powder sintering. Science and Technology of Advanced Materials. 2002;**3**:297-301. DOI: 10.1016/s1468-6996 (02)00030-x

[3] Eustathopoulos N, Nicholas MG, Drevet B. Wettability at High Temperatures. Oxford: Elsevier; 1999. 106 p

[4] Lin Q, Qiu F, Sui R. Characteristics of precursor film in the wetting of Zr-based alloys on ZrC substrate at 1253 K. Thin Solid Films. 2014;**558**: 231-236. DOI: 10.1016/j.tsf.2014.02.074

[5] Be'er A, Lereah Y, Taitelbaum H. Reactive wetting of Hg-Ag system at room temperature. Materials Science and Engineering: A. 2008;**495**:102-107. DOI: 10.1016/j.msea.2007.11.095

[6] de Gennes PG. Wetting: Statics and dynamics. Reviews of Modern Physics. 1985;**57**:827. DOI: 10.1103/RevModPhys.57.827

[7] Li ZK, Ma GF, Fu HM, Sha PF, Zhang B, Zhu ZW, et al. The spreading kinetics and precursor film characteristics of Zr-based alloy melt on W substrate. Materials Letters. 2013;**98**:98-101. DOI: 10.1016/j.matlet.2013.02.005

[8] Lemmens B, Gonzalez Garcia Y, Corlu B, De Strycker J, De Graeve I, Verbeken K. Study of the electrochemical behaviour of aluminized steel. Surface and Coatings Technology. 2014;**260**:34-38. DOI: /10.1016/j.surfcoat.2014.06.064

[9] Gupta SP. Intermetallic compound formation in Fe–Al–Si ternary system: Part I. Materials Characterization. 2002;**49**:269-291. DOI: 10.1016/S1044-5803(03)00006-8

[10] Protsenko P, Terlain A, Traskine V, Eustathopoulos N. The role of intermetallics in wetting in metallic systems. Scripta Materialia. 2001;**45**:1439-1445. DOI: 10.1016/S1359-6462(01)01181-2

[11] Durandet YC, Strezov L, Ebrill N. Formation of Al-Zn-Si coatings on low carbon steel substrates. In: Proceedings of the 4th International Conference on Zinc and Zinc Alloys Coated Steel Sheet (GALVATECH '98). Chiba: The Iron and Steel Institute of Japan; 1998. pp. 147-151

[12] Barin I. Thermochemical Data of Pure Substances. 3rd ed. Weinheim: Wiley-VCH Verlag GmbH; 1995. 702 p

[13] Miller CJ, Pa P. Fluxless aluminum brazing. US Patent: 3373483; 1968

[14] Zhuang H, Lugscheider E. High Temperature Brazing. Beijing: National Defense Industry Press; 1989. 158 p

[15] Perrot P. Aluminum—silico n-titanium. In: Effenberg G, Ilyenko S, editors. Light Metal Systems, Part 4: Selected Systems from Al-Si-Ti to Ni-Si-Ti. Berlin-Heidelberg: Springer-Verlag; 2006. pp. 1-15. DOI: 10.1007/11008514_2

[16] Zhang RF, Sheng SH, Liu BX. Predicting the formation enthalpies of binary intermetallic compounds. Chemical Physics Letters. 2007;**442**: 511-514. DOI: 10.1016/j.cplett.2007.06.031

[17] Ishida T. Spreading kinetics of liquid metals on mild steel. Journal Materials Science and Technolog. 1988;**4**:830-835. DOI: 10.1179/mst.1988.4.9.830

[18] Chen S, Li L, Chen,Y, Liu D. Si diffusion behavior during laser weldingbrazing of Al alloy and Ti alloy with Al-12Si filler wire. Transactions of Nonferrous Metals Society of China. 2010;**20**:64-70. DOI: 10.1016/S1003-6326(09)60098-4

Chapter 5

Influence of Al_2O_3 Processing on the Microtexture and Morphology of Mold Steel: Hydrophilic-to-Hydrophobic Transition

Kelvii Wei Guo

Abstract

The surface of mold steel was processed by the simple Al_2O_3 surface processing, and the influence of processing time on the surface morphology was studied by 3D profilometer and scanning electron microscopy (SEM). Moreover, the wettability of the Al_2O_3 microtextured surfaces of mold steel was also investigated. The results show that the surface morphology of mold steel varies with Al_2O_3 processing time. It reveals that the initial surface without any Al_2O_3 processing treatment behaves as a hydrophilic surface. With the increment of Al_2O_3 processing time, the surface roughness of the processed surface with the microtextures increases correspondingly. At the same time, the wettability of the microtextured surfaces changes from the hydrophilic to the hydrophobic. When Al_2O_3 processing time reaches 60 min, the contact angle reaches its maximum at which the relevant surface roughness is the minimum. It indicates that mold steel with an Al_2O_3 microtextured surface can be a potential application in the mold releasement.

Keywords: surface morphology, Al_2O_3, microtexture, wettability, hydrophilic, hydrophobic, processing, roughness, mold steel

1. Introduction

Conventionally, the surface finishing of metal molds is always done by hand lapping after processing and/or electrical discharge machining in order to attain small surface roughness without microcracks. However, in this process, operator shortcomings cause these manual processing methods to have a number of limitations. Meanwhile, consistency and repeatability are also required. Consequently, the process is extremely time consuming, which leads to high cost [1].

While automated processes suitable for the finishing of closed dies, they are limited in their application. For example, precision machining using a single point diamond tool is slow, requires conditions not readily available in an industrial environment, and is limited to flat surfaces [2–4]. Chemical micromachining and electrochemical micromachining are limited in their application and can be difficult to control [4–9]. The laser has also been widely used as a machine tool to modify the surface of the engineering materials, such as laser surface alloying, laser cladding, surface texturing, laser physical vapor deposition, laser polishing, etc. [2, 10–20].

Ultimately, surface modifications or surface treatments are vitally important for increasing service life of the critical components and devices used for engineering and structural functions. Numerous surface engineering approaches are employed such as thermal, chemical, mechanical, as well as hybrid treatments to improve or vary/change the surface finish.

In this study, the influence of alumina-based surface processing on the micro-texture, morphology, and wetting behavior of mold steel has been investigated. The morphology of initial as well as processed surfaces was investigated as a function of processing time. After being processed, the influence of processing time on the surface morphology of mold steel was studied by 3D profilometer and scanning electron microscope (SEM). The wettability of the processed surface is also investigated.

2. Materials and methods

2.1 Materials

The chemical composition of mold steel tool steel is shown in **Table 1**.

2.2 Methods

The materials were processed into 25 mm × 25 mm × 5 mm slabs and carefully cleaned with acetone and pure ethyl alcohol to remove any contaminants on its surface.

The planetary mono mill "Pulverisette 6" (Made in Germany) was used for surface processing in the stainless steel processing bowl (volume: 500 ml). The process was performed under the vacuum to prevent contaminations for time periods ranging between 15 and 180 min at a processing speed of 250 rpm. All the processed

Element	C	Si	Mn	Cr	W	V	Fe
(wt.%)	0.9	0.3	1.2	0.5	0.5	0.1	Bal.

Table 1.
Chemical composition of mold steel.

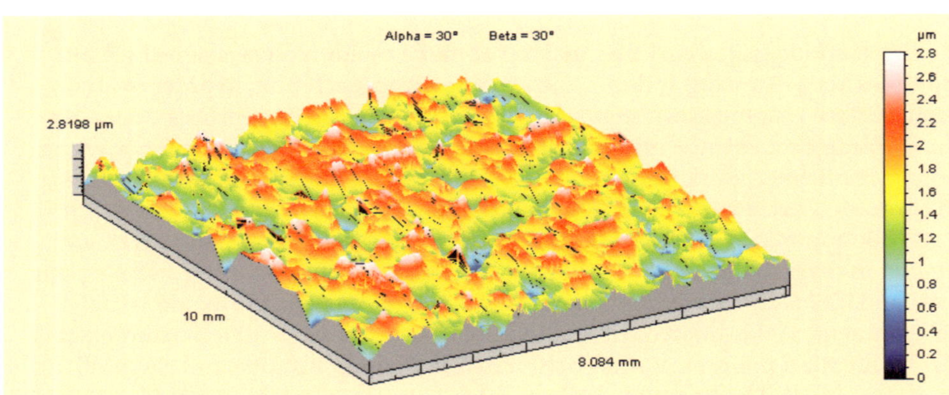

Figure 1.
Morphology of the original surface of mold steel.

Figure 2.
SEM of the initial mold steel surface.

specimens were ultrasonically cleaned in an acetone bath for 10 min with 28–34 Hz frequency and carefully dried. The surface morphology of surfaces was observed by Taylor Hobson profilometer/Talysurf PGI, optical microscope (OM), and scanning electron microscope (SEM) JEOL/JSM-5600. Contact angle (CA) measurement was taken by an advanced contact angle goniometer with DROPimage Advanced (ramé-hart Model 500) attached with a charge-coupled device video camera (with a resolution of 768 × 494 active pixel) and an environmental chamber with temperature control. The volume of the droplet was 10 μL.

Figure 1 shows the Talysurf 3D topography of the original mold steel tool steel specimen, while **Figure 2** shows its corresponding 2D SEM morphology.

3. Results and discussion

3.1 Morphology of the processed surface as a function of processing times

3.1.1 Processing time: 15 min

The morphology of the processed surface after 15 min processing is shown in **Figure 3**, and its related SEM image is expressed in **Figure 4**.

The results indicate that the processed surface is smoother than its initial surface (**Figures 1** and **2**). According to **Figures 3** and **4a**, some small ridges were found distributed on the processed surface. The grooves on the surface had disappeared to a certain extent, together with some smaller island-form ridges (such as labels A and B) due to the shorter processing time. The big ridges cannot be detected and the remained chippings disappear (**Figure 2**), resulted in the smooth processed surface as shown in **Figure 4a**.

Moreover, the magnified parts of **Figure 4a** are shown in **Figure 4b** and **c**, and crack can be found on the processed surface as shown in **Figure 4c** (label C) as a result of the effect of processing balls.

3.1.2 Processing time: 30 min

The morphology of the processed surface after 30 min processing time is shown in **Figure 5**, its corresponding texture is shown in **Figure 6a**, and its magnification

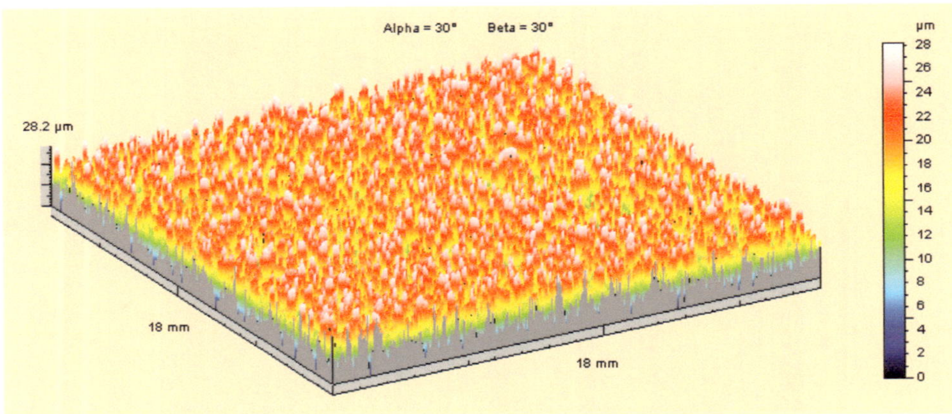

Figure 3.
Morphology of processed surface after 15 min processing time.

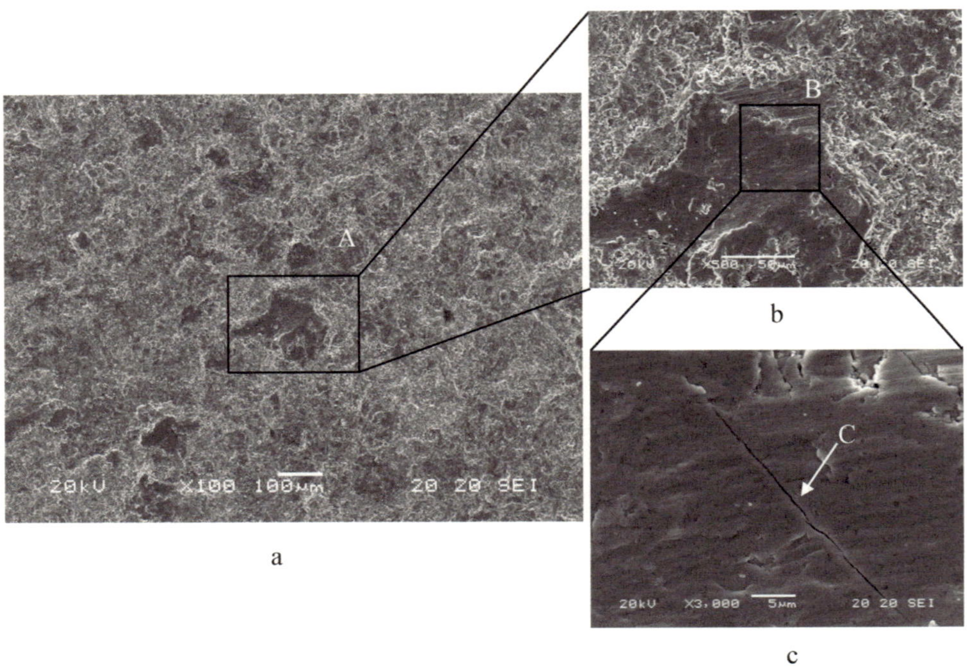

Figure 4.
Processed surface after 15 min processing time.

parts are shown in **Figure 6b** and **c**. Although processing balls remove most high plateaus from the original surface of the specimens (cf. **Figures 2** and **6**), the island-form ridges were found to be still distributed on the processed surface. As expressed in **Figure 6b** and **c**, micropits and cracks can be observed on the surface of the substrate because the balls did not mill sufficiently.

3.1.3 Processing time: 60 min

With the increase of processing time, the processed surface becomes more and more smooth as shown in **Figure 7**, and the formed ridges become smaller and

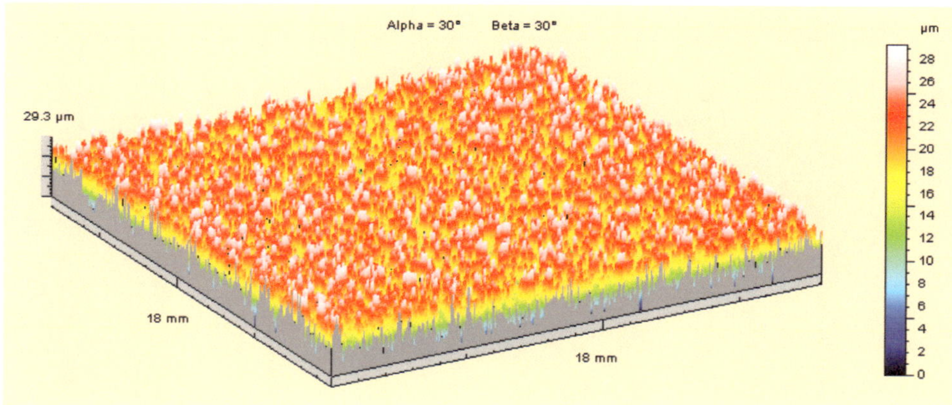

Figure 5.
Morphology of processed surface after 30 min processing time.

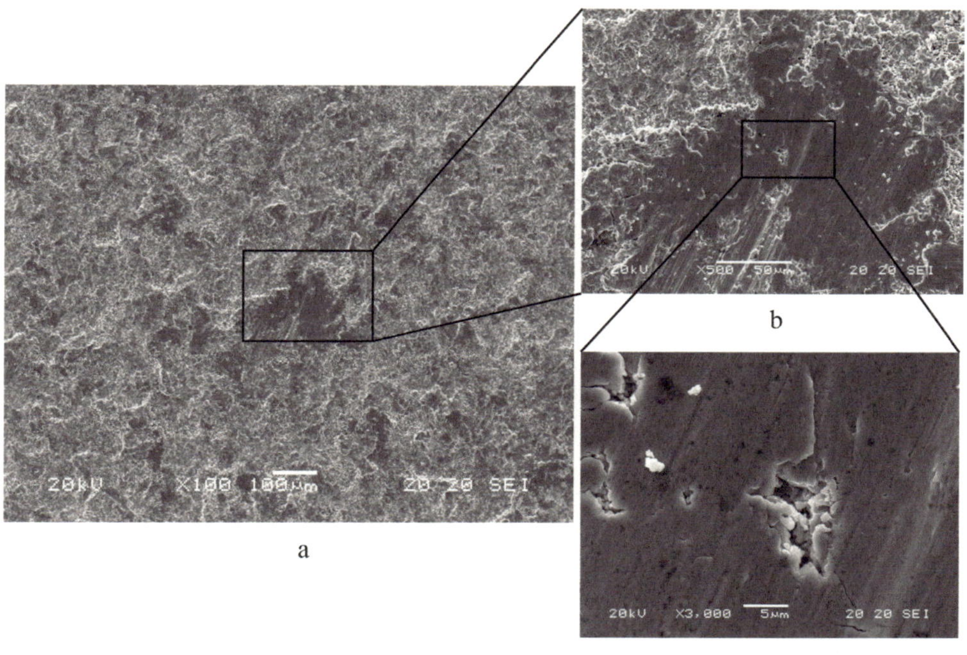

Figure 6.
Processed surface after 30 min processing time.

distribute uniformly as shown in **Figure 8** where the magnifications of **Figure 8a** are shown in **Figure 8b** and **c**. It illustrates the uniform changes of the surface topography after 60 min processing time. The variation of the topography of the surface processed at 60 min is distinct. It is observed that the mold steel surface was processed effectively without any defects (cf. **Figures 4, 6**, and **8**).

Generally, increase in processing time simultaneously improves the properties of the surface substrate. However, too longer processing time will be likely to destroy the surface of the substrate which subsequently changes the surface topography and mechanical properties of the steel specimens.

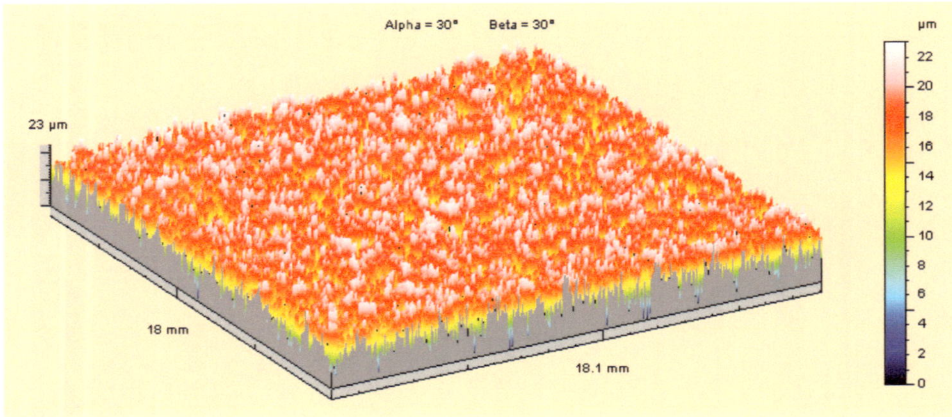

Figure 7.
Morphology of processed surface after 60 min processing time.

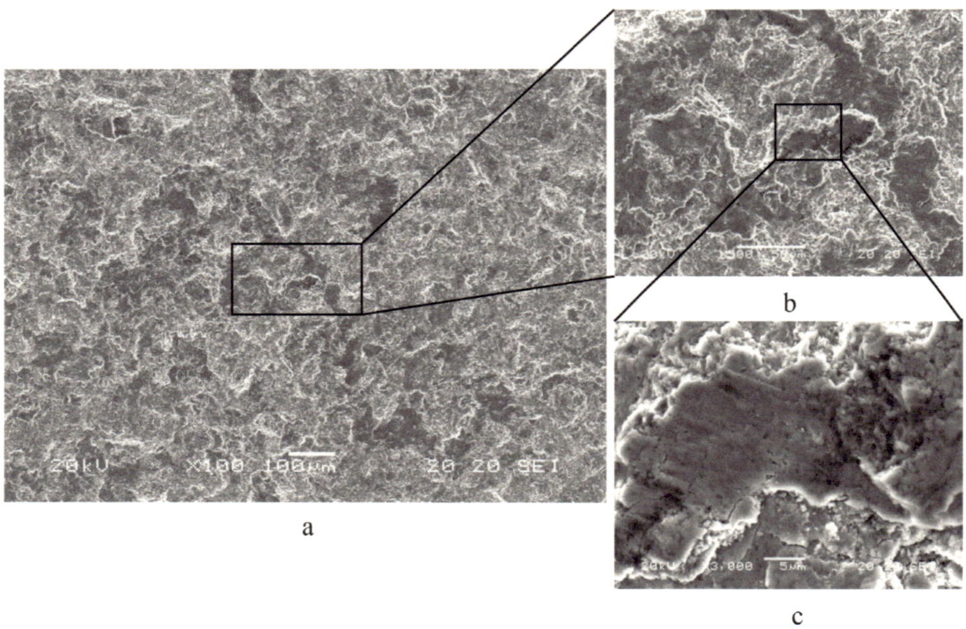

Figure 8.
Processed surface after 60 min processing time.

3.1.4 Processing time: 120 min

When compared with the processed surface of 60 min processing time (**Figure 8**), the processed surfaces (**Figures 9** and **11**) of longer processing time give relatively coarser surface. Also, its SEM image (**Figures 10a** and **12a**) and the further magnified counterparts (**Figures 10b, c** and **12b, c**) show the sign of ridges regrowing bigger, some microparticles aggregating loosely, and microcracks scattering on the processed surface. Such surface topography with scattering of microaggregation and micro ball-like amorphous features (**Figure 10c**) implies that there is some level of change of properties of the mold steel surface. This change is not really anticipated since it is initially expected that the properties of the processed surface topography would be the same as its as-received condition or higher than the initial one.

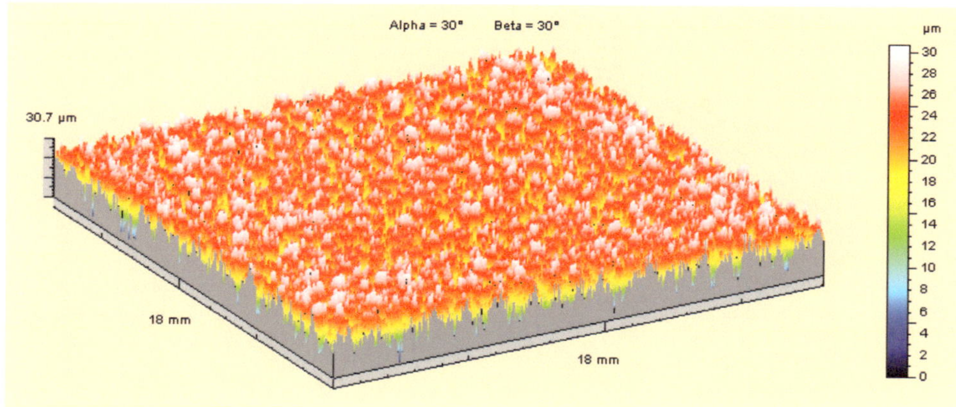

Figure 9.
Morphology of processed surface after 120 min processing time.

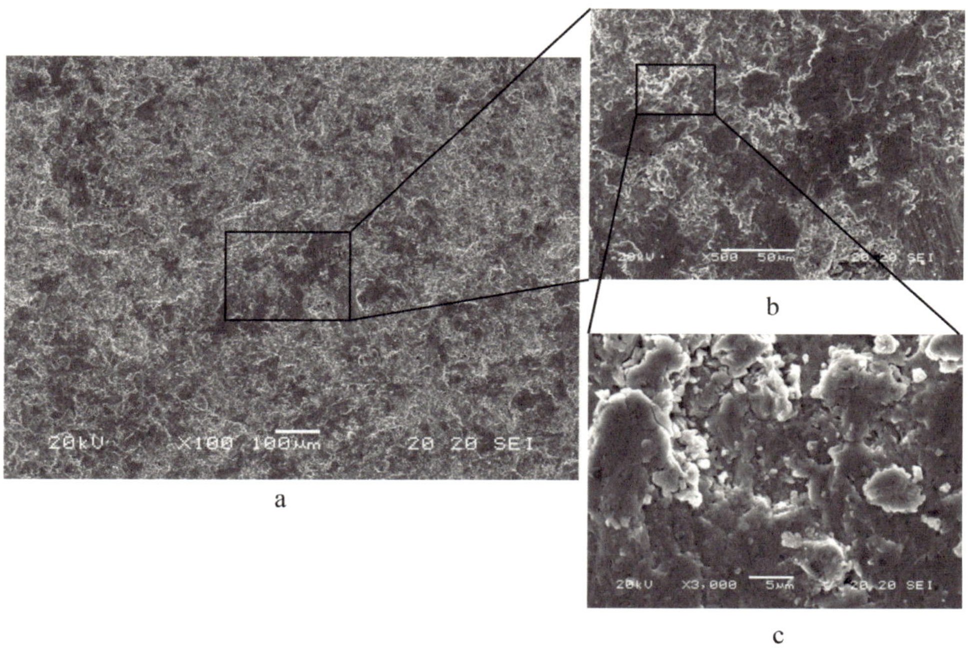

Figure 10.
Processed surface after 120 min processing time.

3.1.5 Processing time: 180 min

A longer processing time (i.e., 180 min), besides the morphology of the processed surface changes obviously (**Figure 11**), relatively more severe cracks appear as shown in **Figure 12b** and **c**, which are likely to change the properties of the initial surface drastically.

3.2 Determination of surface roughness

Figure 13 shows the relationship between the arithmetic mean surface roughness *Ra* and the processing time. Results indicate that the initial increase in processing

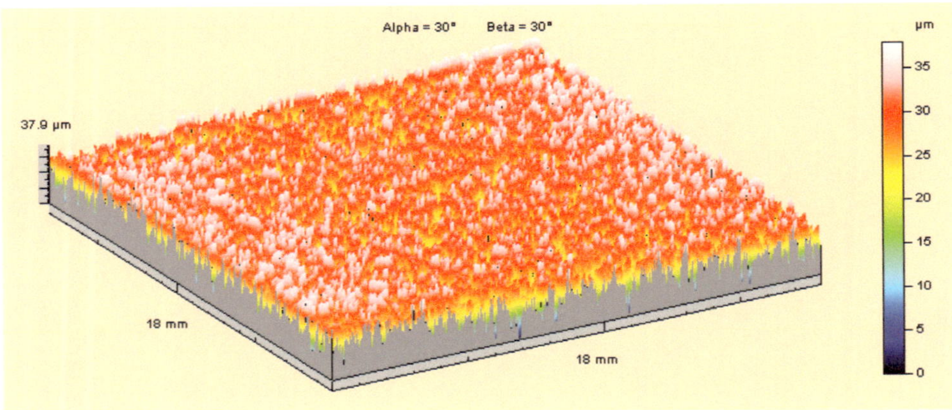

Figure 11.
Morphology of processed surface after 180 min processing time.

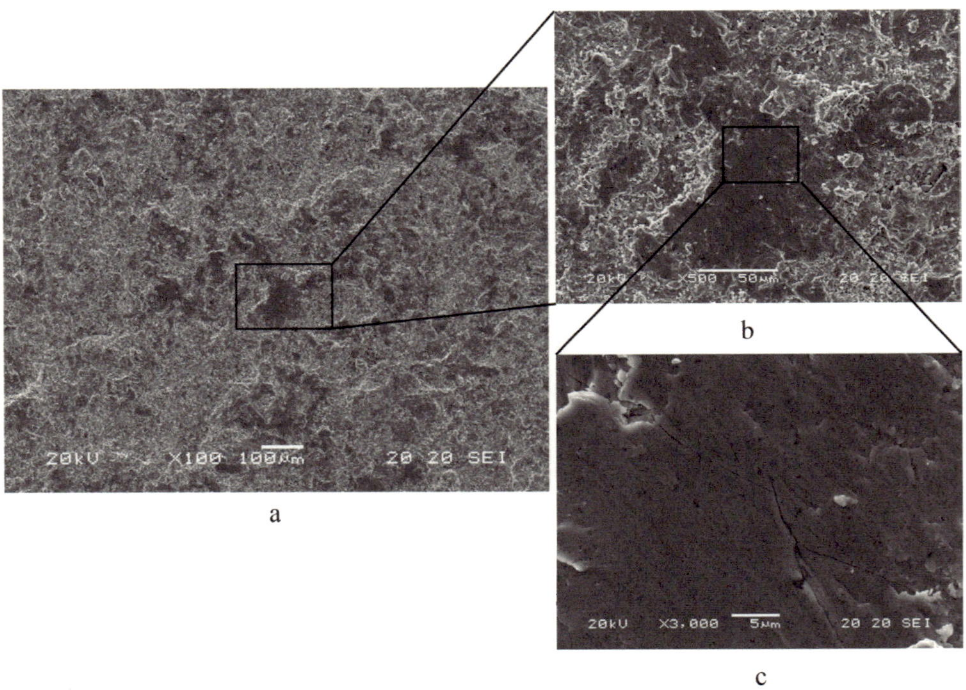

Figure 12.
Processed surface after 180 min processing time.

time accompanies with the increase in surface roughness until the processing time reaches 60 min at which the surface roughness is the minimum. Then, a further increase in the processing time increases the roughness once again which agrees well with morphology as shown in **Figures 3–12**.

3.3 Wettability of milled surface

The variation of CA during the processing process is shown in **Figures 14** and **15**. It is noted that the evolution of CA is related to the processing time. For the initial surface of mold steel, its wettability is hydrophilic. When the processing

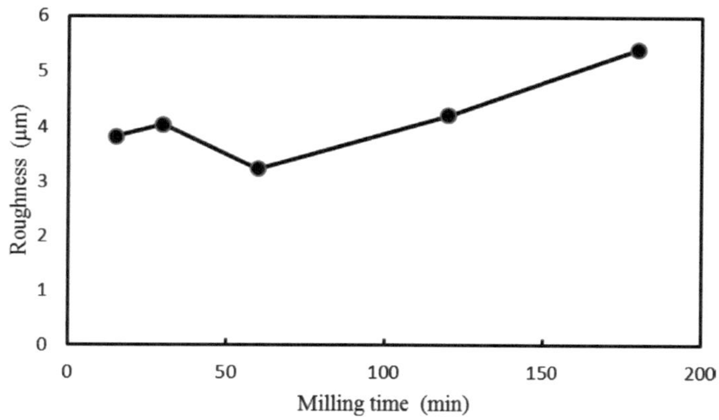

Figure 13.
Correlation between Ra and processing time.

increases, the wettability of the surface varies obviously. The wettability changes from hydrophilic to hydrophobic when the processing time is 60 min, which is an attractive characteristic potential for model release. Moreover, CA increases with the increment of processing time at the early processing period. However, when the surface is processed with longer time, such as 120 and 180 min, the wettability of the surface will be hydrophilic again as shown in **Figure 15**.

It is well known that there is a distinction between the "actual surface" of an interface and the "geometric surface," which is measured in the plane of the interface. At the surface of any real solid, the actual surface will be greater than the geometric surface because of surface roughness. Due to this distinction, the contact angle will be influenced by the roughness. When the surface roughness is considered, the contact angle and droplet profile will change to keep the equilibrium. To evaluate the effect of roughness on surface wettability and calculate the new contact angle of θ' on the rough surface, two different models were proposed by Wenzel and Cassie-Baxter. Both models emphasize on the geometry feature of solid surfaces that acts as a critical factor in determining the wettability.

A system with a droplet placed on a rough solid surface is considered as shown in **Figure 16**, in which the surface texture feature size is much smaller than the droplet, so the influence of liquid weight on an indentation can be neglected when compared to that from surface tension. In traditional theory, whether air can be trapped between liquid and the solid surface is determined by the surface tension of liquid. The liquid intends to exhibit its intrinsic contact angle ($\theta\mathbf{0}$) on the edge of islands. On a hydrophilic substrate ($\theta\mathbf{0}$ < 90°), concave menisci form in the indentations, and the result of liquid surface tension is directed downward and drives the liquid to fill the indentations as much as possible, as shown in **Figure 16a**. On the other hand, for a hydrophobic substrate ($\theta\mathbf{0}$ > 90°), convex menisci form, and the surface tension of liquid is directed upward and pushes the liquid to be suspended on indentations, as shown in **Figure 16b**. Considering of that θ_w is smaller than θ_0 for hydrophilic materials and θ_c is larger than θ_0 for hydrophobic materials, as a result, on an ideal patterned surface, the contact angle would always decrease when $\theta\mathbf{0}$ < 90° and be increased when $\theta\mathbf{0}$ > 90°. Therefore, one can make the option of the processing time with the ideal surface property according to the practical applications.

3.4 Preliminary computations

During the processing process, the mean value of the magnitude of the critical torque $\overline{\tau_{critical}}$ can be expressed as:

$$\overline{\tau_{critical}} = \lambda_\tau d_t \left[\!\left[F_n + 3 f_{adhesion} \right]\!\right] \tag{1}$$

where λ_t is a constant in the range $0.5 < \lambda_t < 1$, d_t is the distance parallel to the plane from the center to one of the asperities in contact, $f_{adhesion}$ is an adhesive force at each contact, and F_n is the normal body forces.

$$\overline{\varnothing_{critical}} = \lambda_\varnothing \frac{2\left(\frac{3|P + f_{adhesion}|}{4E^*\sqrt{R^*}}\right)^{\frac{2}{3}}}{3d_t} \tag{2}$$

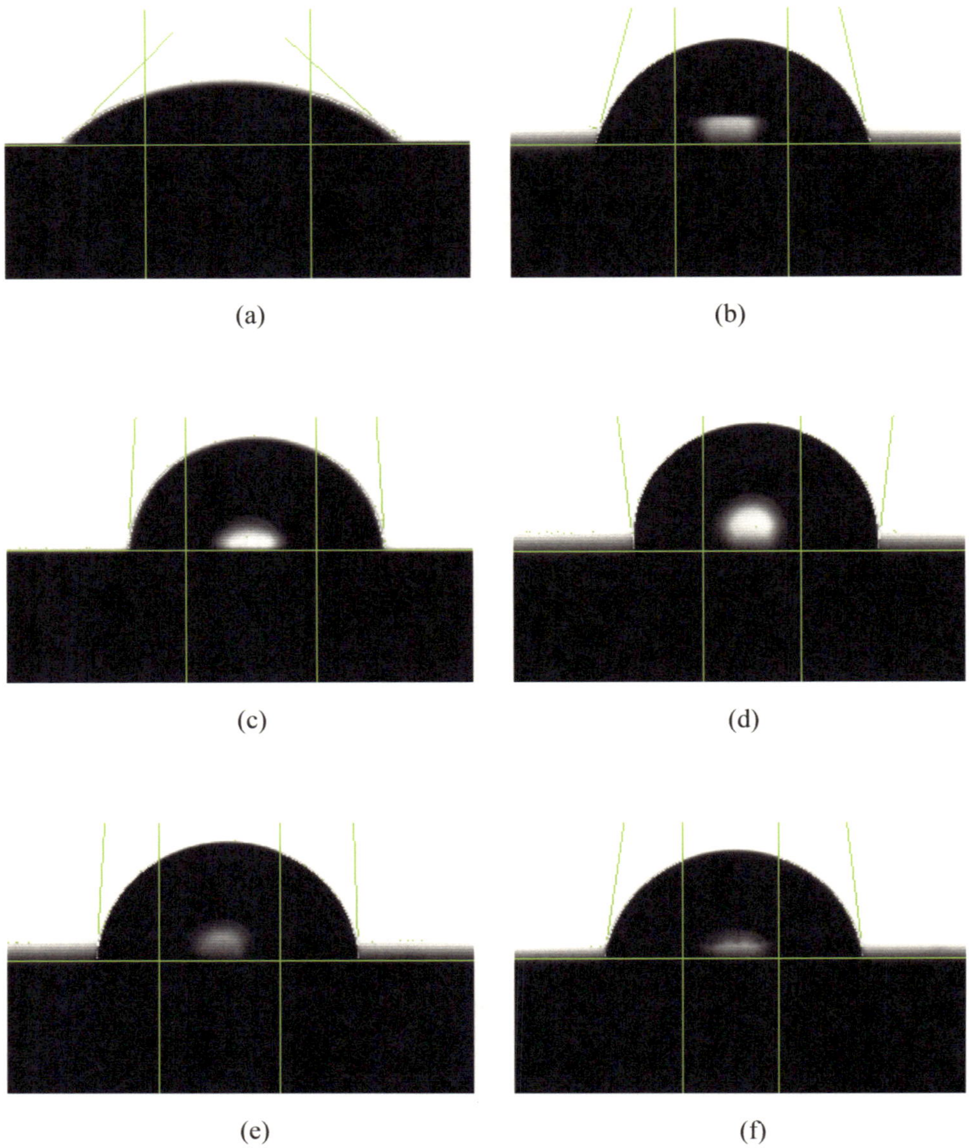

Figure 14.
Variation of contact angle (CA) at different processing times. (a) CA of the initial mold steel surface, (b) CA of the surface at 15 min processing time, (c) CA of the surface at 30 min processing time, (d) CA of the surface at 60 min processing time, (e) CA of the surface at 120 min processing time, and (f) CA of the surface at 180 min processing time.

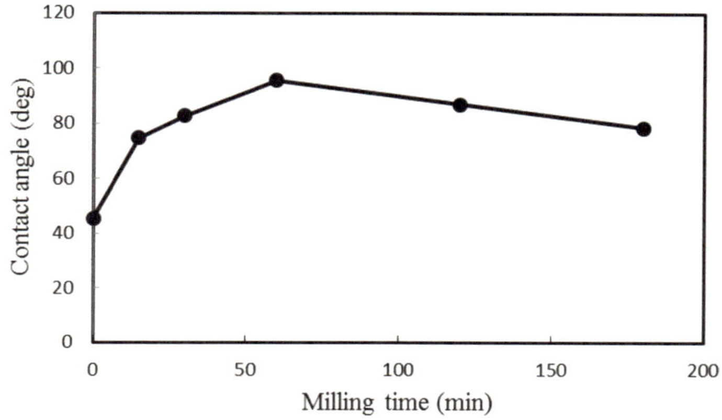

Figure 15.
Correlation between CA and processing time.

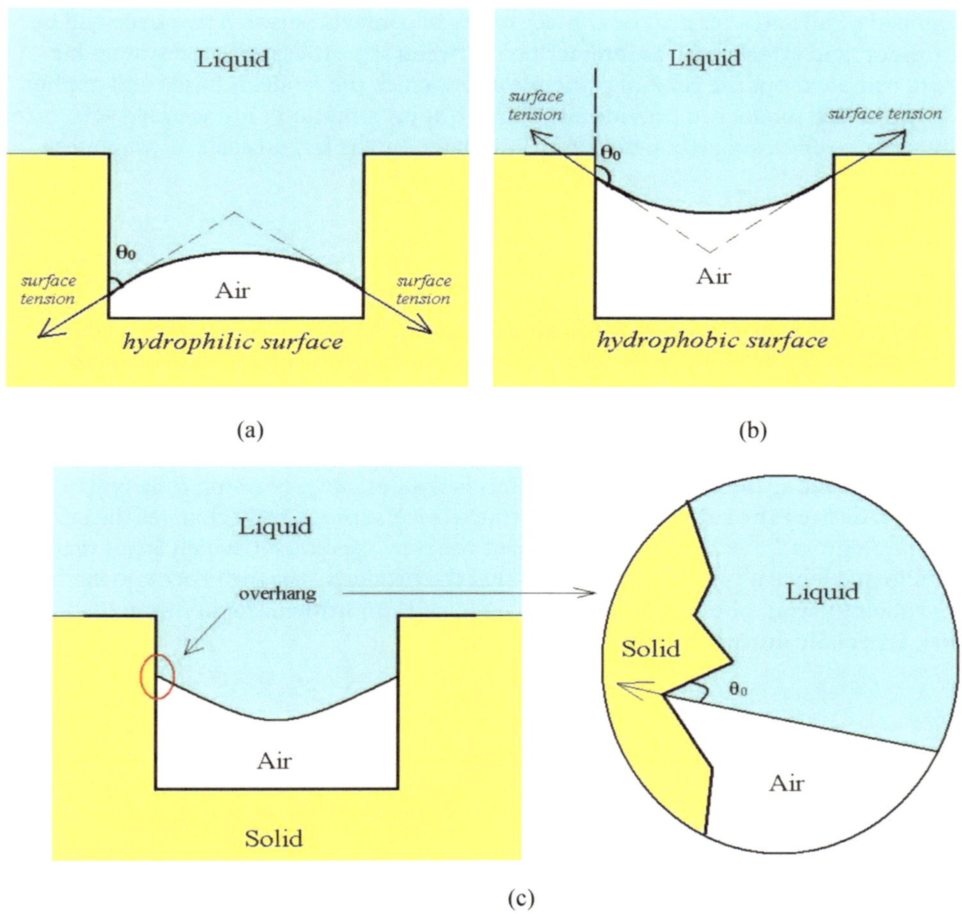

Figure 16.
Schematic representation of the cross section of the droplet-substrate system in indentations. (a) Hydrophilic substrate, (b) hydrophobic substrate, and (c) overhang on a hydrophilic surface with microscopic roughness.

$\overline{\varnothing_{critical}}$ is the critical angle at which critical torque $\overline{\tau_{critical}}$ occurs, $0.63 < \lambda_{\varnothing} < 1$, P is the total load, $R^* = \frac{R_1 R_2}{R_1 + R_2}$ is the reduced radius, and $E^* = \frac{E_1 E_2}{E_1(1 - \nu_2^2) + E_2(1 - \nu_1^2)}$ is the reduced elastic modulus.

$$\overline{\tau_{decelerating}} = -\gamma\omega \qquad (3)$$

where $\gamma = -2\xi\eta_n |d_t|^2 n_{2\pi}$.

$\overline{\tau_{decelerating}}$ is the decelerating torque, ξ is the coefficient ($\xi < 1, or\ \xi \ll 1$), η_n is the damping coefficient, ω is the angular velocity, $n_{2\pi}$ is the number of asperities per revolution, and γ is the adhesive force.

The model consists of expressions for the critical angle and torque at which a ball starts to mill, as well as the rate at which they decelerate. Because of the stochastic nature of surface roughness, it is impossible to accurately reproduce all behavior in a model simple enough to be used in the ball processing phenomena reconstructions. While average processing effect can be accurately replicated, the contact between a real ball and the substrate will not necessarily follow this average due to the details of the geometry. However, in a system of many balls, it is often the case that the average behavior dominates. While variations around the average may have a large effect on the motion of each individual ball, the statistical behavior of a system of many balls will not be significantly changed. The model has been derived for the contact between a ball and a plane. The contact forces in contacts between two balls will be different, and effects such as interactions between asperities make the system far more complex, but the general principle upon which the model is based still applies. The proposed model can provide an adequate approximation of processing effects in the surface processing when torque is dominated by the largest scale of roughness.

4. Conclusions

The morphology of mold steel varies with the different processing times. Increase in processing time simultaneously makes the wettability of processed surface changed from hydrophilic to hydrophobic. Meanwhile, the initial increase in processing time accompanies with the increase in surface roughness until the processing time reaches 60 min at which the surface roughness is the minimum. However, with the increment of processing time, the wettability of the surface will be hydrophilic again. What is more serious is that longer processing time will be likely to destroy the surface of the substrate which subsequently changes the surface morphology and mechanical properties of the steel specimens, which is not really anticipated since it is initially expected that the properties of the processed surface morphology would be the same as its as-received condition or higher than the initial one, especially during the mold releasement.

Author details

Kelvii Wei Guo
Department of Mechanical and Biomedical Engineering, City University of Hong Kong, Kowloon, Hong Kong

*Address all correspondence to: yeelikwok@yahoo.com

References

[1] Maghsoudi K, Jafari R, Momen G, Farzaneh M. Micro-nanostructured polymer surfaces using injection molding: A review. Materials Today Communications. 2017;**13**:126-143

[2] Roberts AG, Krauss G, Kennedy LR. Tool Steels. 5th ed. Ohio: ASM International, Materials Park; 1998

[3] Dehghan-Manshadi A, Bermingham MJ, Dargusch MS, StJohn DH, Qian M. Metal injection moulding of titanium and titanium alloys: Challenges and recent development. Powder Technology. 2017;**319**:289-301

[4] Ryk G, Kligerman Y, Etson I. Experimental investigation of the laser surface texturing for reciprocating automotive components. Tribology Transactions. 2002;**45**(4):444-449

[5] Assarzadeh S, Ghoreishi M. Prediction of root mean square surface roughness in low discharge energy die-sinking EDM process considering the effects of successive discharges and plasma flushing efficiency. Journal of Manufacturing Processes. 2017;**30**:502-515

[6] Shah Mohammadi M, Ghani M, Komeili M, Crawford B, Milani AS. The effect of manufacturing parameters on the surface roughness of glass fibre reinforced polymer moulds. Composites Part B: Engineering. 2017;**125**:39-48

[7] Wojciechowski S, Maruda RW, Nieslony P, Krolczyk GM. Investigation on the edge forces in ball end milling of inclined surfaces. International Journal of Mechanical Sciences. 2016;**119**:360-369

[8] Vanarase A, Aslam R, Oka S, Muzzio F. Effects of mill design and process parameters in milling dry extrudates. Powder Technology. 2015;**278**:84-93

[9] Brito TG, Paiva AP, Ferreira JR, Gomes JHF, Balestrassi PP. A normal boundary intersection approach to multiresponse robust optimization of the surface roughness in end milling process with combined arrays. Precision Engineering. 2014;**38**(3):628-638

[10] Tangwarodomnukun V. Cavity formation and surface modeling of laser milling process under a thin-flowing water layer. Applied Surface Science. 2016;**386**:51-64

[11] Ahmed Obeidi M, McCarthy E, Brabazon D. Methodology of laser processing for precise control of surface micro-topology. Surface and Coatings Technology. 2016;**307**:702-712

[12] Gisario A, Puopolo M, Venettacci S, Veniali F. Improvement of thermally sprayed WC–Co/NiCr coatings by surface laser processing. International Journal of Refractory Metals and Hard Materials. 2015;**52**:123-130

[13] Vilar R, Sharma SP, Almeida A, Cangueiro LT, Oliveira V. Surface morphology and phase transformations of femtosecond laser-processed sapphire. Applied Surface Science. 2014;**288**:313-323

[14] Yilbas BS, Khaled M, Abu-Dheir N, Aqeeli N, Furquan SZ. Laser texturing of alumina surface for improved hydrophobicity. Applied Surface Science. 2013;**286**:161-170

[15] Guo KW, Tam HY. Study on polishing DF2 (AISI O1) steel by Nd:YAG laser. Journal of Materials Science. 2012, 2012;**1**(1):54-77. DOI: 10.5539/jmsr.v1n1p54

[16] Guo W, Hua M, Tse PW-T, Mok ACK. Process parameters selection for laser polishing DF2 (AISI O1) by Nd:YAG pulsed laser using orthogonal

design. The International Journal of Advanced Manufacturing Technology. 2012;**59**(9-12):1009-1023. DOI: 10.1007/s00170-0113558-1

[17] Kelvii Wei GUO. Effect of polishing parameters on morphology of DF2 (AISI-O1) steel surface polished by Nd:YAG laser. Surface Engineering. 2009;**25**(3):187-195

[18] Wei GUO. Effect of irradiation parameters on morphology of polishing DF2 (AISI-O1) surface by Nd:YAG laser. Advances in Materials Science and Engineering. 2007;**3**:5-9. DOI: 10.1155/2007/51316

[19] Lugomer S. Laser Technology-Laser Driven Processes. Englewood Cliffs, NJ, USA: Prentice Hall Inc; 1990. pp. 419-439

[20] Steen WM. Laser Material Processing. 3rd ed. London: Springer-Verlag; 2003

Chapter 6

Interfacial Phenomena in the Synthesis Process of Barium Sulfate Particles Precipitated in a Lobed Inner Cylinder Taylor-Couette Flow Reactor: Effects of Fluid Dynamics

Lu Liu, Guang Li, Xiaogang Yang, Xiani Huang and Chenyang Xue

Abstract

Three different kinds of morphology with various sizes of barium sulfate particles were produced by reactive precipitation in a Taylor-Couette flow reactor. It is found that particle morphology transition is strongly related to the hydrodynamics in the reactor, clearly indicating an interfacial interaction between feed solutions and aggregated particles. At low concentration, particle morphology transition is observed at the onset of turbulent Taylor-Couette flow. Such morphology transition also appears at the onset of turbulent Taylor vortex flow at high concentration. Based on different transition status, supersaturation is found to play an important role in nucleation and growth processes. In addition, it is revealed that the synthesized particle reduces its size as the consequence of the transition in particle morphology, indicating the effect of variation of the feeding rates. Experimental results have confirmed that controllable synthesis of barium sulfate particles with a particular morphology can be achieved through suitable selection of the controlling parameters such as the rotational speed of inner cylinder of Taylor-Couette flow reactor, reactant feeding rate and supersaturation ratio.

Keywords: barium sulfate, Taylor-Couette reactor, particle morphology, flow pattern, supersaturation, agglomeration

1. Introduction

Precipitation is a traditional industrial process to produce solid particles. It has been used in many applications. Typical examples can be found in fabrication of pigments, ceramics, pharmaceuticals and bio-chemicals among others. For various purposes, the requirements of particle property are different. As particle size and morphology play important roles in determining particle property [1], many previous studies have paid attention to realization of particle size distribution control and to acquirement of a particular morphology [2–4]. As many factors can affect final particle property, such as reactant concentration, feeding modes, additives, and reactor system, their interrelationships are complicated so that many precipitation mechanisms have been proposed [5]. However, none of these proposed precipitation mechanisms can fully address the dynamic processes involved for synthesis of the particles given a particular reactor system. The present work focuses on the effects of three main parameters, rotational speed, feeding rate and supersaturation when employing a Taylor-Couette reactor for synthesis of particles on the change of particle morphology, in particular investigating barium sulfate system. Even though the synthesis of barium sulfate particles has been extensively studied [6–8], controllable synthesis of the particles is still not fully understood, which requires further investigations.

Barium sulfate is a sparingly soluble salt, whose crystallization kinetics has been widely studied. Also, its precipitation system from aqueous barium chloride and sodium sulfate is available according to a series of studies by Nielsen [9–11]. In terms of particle morphology, a number of studies have been conducted to investigate different aspects of the potential possible factors. Many morphologies were observed in connection with the synthesis processes, involving tabular particles (or flat particles), dendritic particles, round-shaped particles (or spherical particles), rhombic particles and rice-shaped particles. Barresi et al. [12] produced barium sulfate particles in a continuous Couette reactor, and they observed dendritic tabular crystals and tabular crystals with pyramidal by changing supersaturation, rotational speed and internal diameter of inner cylinder at the same time. They suggested that supersaturation was the dominant factor in determining particle morphology and size. However, they failed to reveal the effect of each individual variable. Marchisio et al. [13] found a series of morphology change from tabular particles, then dendritic particles to rounded-shaped particles with increase of sodium sulfate concentration. They indicated that the excess of species had a stronger effect on particle morphology due to the preferential absorption of the excess ion. The effect of additives, such as EDTA, phosphate, and lanthanum, has been widely investigated experimentally in their study. From the results as reported by Li et al. [14], various morphologies of barium sulfate particles were exhibited when treatment is done by using polyacrylic acid, such as ellipsoids, monodisperse spheres and rose-like aggregates. They contributed the occurrence of these morphologies to the interactions between carboxyl groups of the additive and inorganic ions. However, the mechanism of randomly coiled conformation of the additive was not clearly illustrated, which results in controllable preparation of each morphology in industrial application to be still questionable. By means of barium sulfate precipitation system, Baldyga and his co-workers [15–17] have experimentally conducted a number of studies using the mixing tank and numerically explored the mixing behavior, aiming at fundamentally reveal the mechanism involved. During this process, they found that various parameters, including feed volume ratio, stirred speed, feeding time, volume ratio, initial concentration and stoichiometric ratio have impacts on particle size distribution and morphology. Also, they proposed a mixing-precipitation model which reasonably described the interaction between the micro-mixing timescale and reaction timescale, consistent with the experimental results. Pagliolico et al. [18] have also illustrated the effect of the mixing on the production of various morphologies of barium sulfate, including dendritic particles, tabular particles, rose aggregates. They proposed shape factors for qualitative description of different morphologies. It can be seen that the abovementioned studies have focused on the effect of both solution conditions (typically excessive species, additives or supersaturation) and operating parameters (typically stirred speed, feeding point location, or addition time) on particle properties. However, most of deductions in these studies only focused on the change of particle size with the parameters as mentioned while the obtained results just showed the morphology for a particular condition. Little attention has been paid on how the morphology transits with respect to the variations of operating parameters and the correlations between the morphology and these parameters, in particular flow field environment and initial solution conditions.

Taylor-Couette reactors have been employed to prepare different kinds of particles due to its advantages [3, 6, 19]. As no stirrer is involved, the breakage of particles due to the elastic collision can be avoided. Taylor-Couette flow reactor consists of two co-axial cylinders, whose internal cylinder is rotating while the external cylinder is usually kept stationary. The typical characteristics of hydrodynamics of the reactor are a narrow shear rate distribution and relatively uniform kinetic energy dissipation. An increase in the rotational speed can effectively enhance the mixing efficiency. Jung et al. [3] employed a Taylor-Couette reactor to prepare calcium carbonate particles in a gas–liquid system. After comparing the effects of species excess and shear stress, they suggested that the excessive species have a stronger effect on synthesized particle size and morphology while the effect of flow dynamics on the synthesis can be negligible. It should be pointed out, however, that their analysis focused on the aspect of mass transfer, and overlooked the effect of shear rate variations. Mayra and Kim [20] and Thai et al. [19] have systematically conducted a series of studies on the synthesis of Ni-rich hydroxide crystals, a kind of cathode material for lithium ion battery, by using the Taylor-Couette crystallizer. They have clearly demonstrated that under the condition that the Taylor-Couette reactor operates in the range of the critical Reynolds number based on different gap sizes (Re_c = 128.5–219.4), the formed Taylor vortices promote the mixing and enhance the mixing residual time, which yields the uniform agglomerate particles with high tap density. This clearly indicates that the hydrodynamics of the Taylor-Couette flow reactor will have a significant impact on the synthesis of particles and the particle crystals growth.

Even though reactive precipitation processes are very fast, Barresi et al. [12] have demonstrated that mixing, especially micro-mixing, has a significant influence on the precipitation process itself. The mixing in mixing vessels that is caused by shear can be characterized by macro-mixing, meso-mixing, and micro-mixing. According to engulfment-deformation-diffusion model (EDD) as proposed by Baldyga and Bourne [21], engulfment due to the micro-mixing generates the local supersaturation and dilution of species. Such local supersaturation is the driving force for crystallization. Macro-mixing occurs on the scale of the reactor, functioning to convey the reactive solutions throughout the entire available space of the reactor [22]. Macro-mixing provides the environment for the following mixing processes, and affects the distribution of supersaturation. Meso-mixing refers to the exchange of fresh feed and its surroundings on a coarse scale, lager than Kolmogorov scale, but smaller than integral scale of turbulence. It can influence micro-mixing by changing the local environment [23]. Micro-mixing brings the fluid elements into contact, followed by molecular diffusion. It controls the generation of supersaturation.

The aim of the present study is to investigate the morphology transition process of barium sulfate precipitated in a Taylor-Couette reactor with a lobed internal cylinder [24]. The effects of three parameters, rotational speed, feeding rate, and supersaturation, on the synthesized particle morphology are assessed and the correlations are obtained. In order to examine the effect of hydrodynamics characterized by shear rate which closely associates with the rotational speed, a wide range of rotational speeds were chosen varying from 25 to 1000 rpm. The interrelationship of shear rate and final particle morphology is acquired. Then, results will be discussed from the perspective of micro-mixing, as precipitation process is triggered by supersaturation, whereas micro-mixing controls the generation of local supersaturation. The experimental data obtained from current work can be used for the further validation of multiphase micro-mixing model.

2. Wettability and interfacial phenomena—implications for barium sulfate particle precipitated using a lobed inner cylinder Taylor-Couette flow reactor

The role of surface phenomena and contacts can play significant roles in the transitions of particle morphologies during precipitation. From the preceding section, it can be seen that the formation of different types of particles is related to interfacial phenomenon between liquid and particles. The effects of fluid behavior on particles are embodied mainly from two aspects. On one hand, fluid flow in the reactor contributes to the interfacial ion concentration distribution, thus affecting the mass transfer to the aggregated crystal nuclei for formation of barium sulfate particles. Concentration gradient between nuclei surface (so referred to as equilibrium concentration) and bulk solution (so called bulk concentration) is the driving force for particle growth. During this process, in spite of molecular diffusion, convection dispersion resulted from fluid flow has a dominant effects. Many previous studies (e.g., [25–27]) have adopted Sherwood number, defined as the ratio of convective mass transfer to diffusive mass transport, to characterize the mass transfer contributed from turbulence on fluid-particle interface for particle size ranging from macro-particles to micro-particles. This contribution comes from eddy fluctuation, thus indicating that the shear caused by turbulent eddies may play an important role in the formation of particles in the precipitation. Armenante and Kiwan [28] proposed the correlation with Sherwood number for estimation of the mass transfer coefficient for description of the dissolution process of AgCl crystal. They have reaffirmed the effect of turbulence by applying different power input to the mixing tank reactor, and obtained an improved correlation between the Sherwood number and Reynolds number. Furthermore, Jung et al. [3] have obtained different calcium carbonate morphologies by changing the wall shear stress of the Taylor-Couette flow reactor used in their study. They interpreted this phenomenon from the point of view that ion adsorption takes place on the particle surface while the concentration gradient of adsorbed ions is also caused by shear stress. Thus, interfacial turbulence induced shear has an impact on aggregation. Bubakova et al. [29] proposed an equation to describe the relationship between aggregate size and average shear rate of turbulent eddies formed in the mixing process. Similarly, for the process of Ni-rich hydroxide preparation, Mayra and Kim [20] have also confirmed the parameters that are strongly associated with the fluid shear can have significant influence on agglomeration process. They have indicated that individual crystals can stack together to form irregular aggregates with fluid flow (mainly characterized by turbulent eddy motion), and then shear force generated by turbulent eddies can facilitate these aggregates to bind together to form regular agglomerates. Thus, it may be concluded that interfacial ion concentration and interfacial fluid shear between the solution and particles have significant influences on particle growth and aggregation.

3. Theoretical modeling

3.1 Precipitation process dynamical model

The chemical reaction between aqueous solutions of barium chloride and sodium sulfate obeys the following equation,

$$BaCl_{2(aq)} + Na_2SO_{4(aq)} \rightarrow BaSO_{4(s)} + 2NaCl_{(aq)} \tag{1}$$

The precipitation process of barium sulfate involves four main steps, nucleation, growth, agglomeration and breakup. The driving force of nucleation is supersaturation, which has effects on subsequent processes. Supersaturation σ is defined as the ratio of molar chemical potential φ (the molar Gibbs free energy) in supersaturated solution to the potential in saturated solution [23], being expressed by the activities of ions, a,

$$\sigma = \frac{\varphi - \varphi_{eq}}{RT} = ln\frac{a}{a_{eq}} \tag{2}$$

where a_{eq} is the activity in equilibrium, R is gas constant, and T is absolute temperature. The relationship between ion activity and concentration can also be described as follows:

$$a = \gamma_{\pm}c \tag{3}$$

where $\gamma_{\pm}$ is the activity coefficient. Estimation of $\gamma_{\pm}$ has been proposed by Debye and Huckel [30], and the modified equation is usually adopted, given by

$$lg\gamma_{\pm} = -0.511z_{+}z_{-}\left(\frac{\sqrt{I}}{1+\sqrt{I}} - 0.2\sqrt{I}\right) \tag{4}$$

where z is the charge number, and I is the ionic strength, calculated by the following equation

$$I = \frac{1}{2}\sum_{i=1}^{n} c_i z_i^2 \tag{5}$$

where c_i is the molar concentration of species i. Since supersaturation is often expressed as the concentration difference Δc with saturation ratio S defined as the ratio of dissolved concentration to the equilibrium solubility and Sa as a relative saturation ratio, then

$$\Delta c = a - a_{eq} \tag{6}$$

$$S = \frac{a}{a_{eq}} = \sqrt{\frac{a_{Ba^{2+}}a_{SO_4^{2-}}}{K_{SP}}} \tag{7}$$

$$S_a = S - 1 \tag{8}$$

where K_{sp} is the solubility product of barium and sulfate ions in the equilibrium state ($\sim 1.1 \times 10^{-10}$ mol^2/L^2 at 20°C). In the present work, we used S to describe the supersaturation of the system for the following calculation.

During precipitation, new particles are created by nucleation. Molecules of the reactants are combined to generate embryos when the energy required excesses energy barrier ΔG. During this process, there is an induction period which is the necessary time to form a characteristic, significant number of nuclei. This time can also be regarded as the time elapsed from the mixing of reactants to the appearance of nuclei [31], an empirical expression for barium sulfate reaction [32] can be used to estimate t_N

$$lgt_N = 15.5lg^{-2}S - 4.2 \tag{9}$$

Nucleation rate can be described using the following equation

$$J = k_N \Delta c^n \tag{10}$$

where n is the kinetic order of nucleation. In barium sulfate system, for $\Delta c < 0.01$ mol/L, $k_N = 6 \times 10^{14}$, $n = 1.775$; for $\Delta c > 0.01$ mol/L, $k_N = 2.53 \times 10^{39}$, $n = 15$ [15].

Once nuclei have been formed, they start to grow to a particular shape and size. This process can be integrated into two steps: the mass transfer to the solid–solution interface, and the surface integration of growth units diffused into the crystal lattice. The total growth rate given by

$$B = k_g(S - 1)^2 \tag{11}$$

where k_g is growth coefficient. At a temperature of 20°C, $k_g = 4 \times 10^{-11}$ m/s.

3.2 Taylor-Couette vortex flow

Flow pattern in a Taylor-Couette flow reactor can be characterized using the Reynolds number, defined by

$$\mathrm{Re} = \frac{\omega_i r_i \delta_{eq}}{\nu} \tag{12}$$

where ω_i is the rotational speed of the inner cylinder, r_i is the equivalent radius of inner cylinder, estimated based on the ratio of four times of the inner cylinder cross-section area to the wetted perimeter of the cylinder, δ_{eq} is the equivalent gap between the inner cylinder and outer cylinder and ν is the kinematic viscosity of suspension, the ratio of the dynamic viscosity to the density of the fluid. The dynamic viscosity can be measured using a viscometer. For a fixed outer cylinder, with the increase of the rotational speed of inner cylinder exceeding a critical Reynolds number, the circular Couette flow becomes unstable, and Taylor vortices appear in pairs. Taylor vortex flow is a kind of secondary flow, superimposed on the laminar Couette flow. The velocity profiles of Couette flow and Taylor vortex are shown in **Figure 1** respectively. Radial and axial velocities manifest once Taylor vortices appear.

At low rotational speed, the driven flow can be considered as laminar Couette flow, and the fluid velocity is mainly determined by azimuthal velocity, as the highest radial and axial velocities are less than 15% of azimuthal velocity [33]. The azimuthal velocity profile in the gap for incompressible fluid takes the form of

$$u_\theta = Ar + \frac{B}{r} \tag{13}$$

where u_θ is the azimuthal velocity. A and B depend on the radius ratio $\eta = r_i/r_o$ and the cylinder angular velocity ratio $\zeta = \omega_o/\omega_i$.

$$A = \omega_i \frac{\zeta - \eta^2}{1 - \eta^2},\ B = \omega_i \frac{r_i^2(1 - \zeta)}{1 - \eta^2} \tag{14}$$

Then, further increase in rotational speed leads to series of instabilities occurs, changing from laminar flow including Taylor vortex flow, wavy Taylor vortex flow, modulated wavy Taylor vortex flow, to turbulent flow including turbulent Taylor

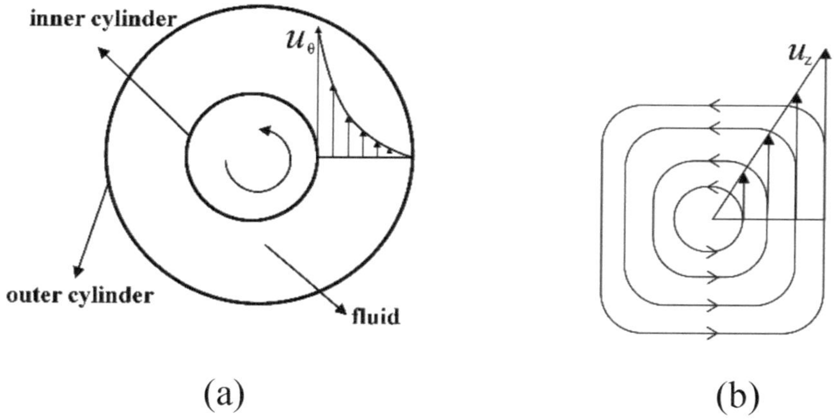

Figure 1.
(a) Azimuthal velocity profiles of circular Couette flow, and (b) axial velocity of Taylor vortex.

vortex flow and fully developed turbulent Taylor flow. These flow patterns are influenced by the use of different aspect ratio and radius ratio. DiPrima et al. [34] have summarized the effects of a series of radius ratio on the flow patterns and some previous studies have investigated the transition points of these patterns [35–37]. At high rotational speed, radial and axial velocities cannot be ignored due to the enhanced effect of Taylor vortices. The mean velocity can be approximately estimated by.

$$\overline{u} = \frac{2\omega_i r_i^2}{3\left(r_o^2 - r_i^2\right)}\left(r_i^3 + 2r_o^3 - 3r_i r_o^2\right) \tag{15}$$

4. Experimental

In the current work, the reactive precipitation of barium sulfate particles was realized with aqueous barium chloride and aqueous sodium sulfate. Barium chloride ($BaCl_2{\cdot}2H_2O$) and sodium sulfate (Na_2SO_4) (analytical grade) were prepared with deionized water, respectively. Sodium pyrophosphate was chosen as the dispersant, added into sodium sulfate, which corresponds to a 10% (w/w) theoretical yield of barium sulfate products. In order to remove solid impurities, all solutions were filtered with filter prior to the storage in containers.

The schematic diagram of experimental setup is shown in **Figure 2(a)**. Experiments were carried out in a Taylor-Couette reactor. It consists of two coaxial cylinders, where the plexiglass outer cylinder keeps static and the aluminum alloy inner cylinder rotates. The dimensions of the Taylor-Couette reactor employed in this study are described in **Table 1**, where the radius ratio is $\eta = r_i/r_o = 40.19/50.00 = 0.80$ and the aspect ratio is $\Gamma = L/\delta_{eq} = 300.00/9.81 = 30.58$. The total volume that can be used for reaction is 922 mL. According to Soos et al.'s study [38], the use of lobed inner cylinder can eliminate or reduce low velocity or shear rate regions, thus improving mixing and mitigating the segregation of particles with different size and density. Therefore, a lobed inner cylinder was chosen for current study. The cross-section profile of both cylinders is shown in **Figure 2(b)**. Due to the feature of lobed inner cylinder, the radius of the inner cylinder will change along the circumferential direction of the cylinder and an equivalent radius based on Eq. (16) is used for evaluation of the gap between the inner and outer cylinders, i.e.,

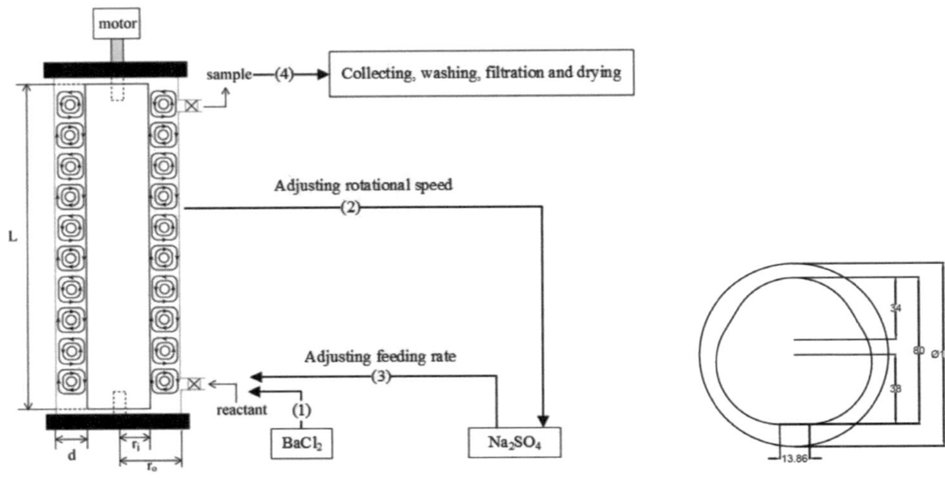

(a) (b)

Figure 2.
(a) Schematic diagram of experimental setup for barium sulfate reactive precipitation in a Taylor-Couette flow reactor with the lobed inner cylinder; and (b) cross-section profile of both outer and inner cylinders.

Equivalent radius of lobed inner cylinder, r_i (mm)	40.19
Radius of outer cylinder, r_o (mm)	50.00
Gap size, δ_{eq} (mm)	9.81
Reactor length, L (mm)	300.00

Table 1.
Dimensions of Taylor-Couette reactor.

$$\delta_{eq} = \frac{2A_i}{R_i} \tag{16}$$

where A_i is the cross sectional area of the inner cylinder, and R_i is the perimeter.

All experiments were conducted at room temperature of 20°C. Barium chloride served as the bulk solution, which was fed into the reactor firstly at the bottom of the reactor before the precipitation reaction takes place by peristaltic pump. The rotational speed of the inner cylinder was controlled by a servo motor, and feeding rate was well-controlled by a constant flow peristaltic pump. They have been carefully calibrated before the experiments. After the rotational speed was adjusted to a particular value, the flow was gradually developed to become steady and the feeding rate of tank solution, sodium sulfate was also adjusted to a required value and was then pumped into the reactor to induce the reaction.

Three parameters were involved in these experiments, including rotational speed, feeding rate, and supersaturation. Two initial concentrations (i.e., two supersaturations) of reactants were prepared at 0.1 and 1 mol/L, respectively. At low concentration, rotational speed had a wide range from 25 to 1000 rpm, and feeding rate changed from 5 to 80 mL/min. In total, 49 runs were carried out. At high concentration, rotational speed varied from 50 to 1000 rpm, and feeding rate changed from 5 to 60 mL/min. There were 32 runs were carried out. All parameters and their values have been summarized in **Table 2**. In addition, in order to ensure the repeatability of the experiments, some of the experiments were repeated at least two times.

	c_{BaCl_2} (M)	$c_{Na_2SO_4}$ (M)	Rotational speed (rpm)	Feeding rate (mL/min)
			25	5, 10, 15, 20, 40, 60, 80
			50	5, 10, 15, 20, 40, 60, 80
			100	5, 10, 15, 20, 40, 60, 80
Low concentration	0.1	0.1	300	5, 10, 15, 20, 40, 60, 80
			600	5, 10, 15, 20, 40, 60, 80
			800	5, 10, 15, 20, 40, 60, 80
			1000	5, 10, 15, 20, 40, 60, 80
			50	5, 10, 40, 60
			100	5, 10, 40, 60
			200	5, 10, 40, 60
High concentration	1.0	1.0	300	5, 10, 40, 60
			400	5, 10, 40, 60
			600	5, 10, 40, 60
			800	5, 10, 40, 60
			1000	5, 10, 40, 60

Table 2.
Operating conditions.

During each experiment, suspension samples were taken continuously from the outlet located at the top of the reactor, and at the end of each run they were quickly diluted with deionized water to quench the reaction. After conducting each experiment, the reactor was cleaned by diluted hydrochloric acid to remove barium sulfate particles that have stuck on the wall of the reactor, then rinsed twice with deionized water. Collected samples were washed three times to sufficiently dissolve sodium chloride co-produced in the reaction and other excessive ions. Products were then filtered with mixed cellulose ester microporous membrane filter (pore size 0.22 μm), dried in the oven at 100°C for 24 hours. For even smaller particles, a 0.1 μm pore filter paper was used. Selected samples were then analyzed by SEM (Scanning Electron Microscope). Before taking the SEM observation, the samples were slightly ground. In order to increase the electrical conductivity, silicon wafer was stuck onto carbon tape. Barium sulfate particles were dispersed into ethanol, then dropped on the surface of silicon wafer. Prepared samples were dried in the oven about 10 minutes before conducting SEM observation. Particle size was measured and characterized by the software, Image J, which should establish a coordinate system based on the scale in SEM images. For each reaction condition, at least three images were selected, and 200 particles were measured in each image.

5. Results and discussion

As it has been mentioned before, final particle properties are influenced by many factors. With no additional additives in the system, the properties of particles are directly influenced by the surroundings where the crystals create and grow. Feeding rate affects the local amount of the fresh reactant fed to the system per unit time, which reflects the meso-mixing efficiency in the Taylor-Couette reactor. Rotational speed controls the dispersion degree of fresh reactant, bringing the unpremixed reactants into contact at a molecular level. It reflects the micro-mixing efficiency of the system. Reactant concentration is the other factor that affects the formation of supersaturation of solution, resulting in the supersaturation being the driving force for the local nuclei formation. Thus, it can be suggested that particle morphology will be controlled by at least these three parameters.

The formation of different morphologies results from the mixing behavior throughout the whole nucleation process, while mixing is caused by fluid shear. Fluid shear generating velocity gradient can re-disperse species, and create eddies. Thus, shear has significant effects on fluid behavior, such as mixing and mass transfer. For a laminar flow, shear rate in the gap can be estimated from the derivative of Eq. (13)

$$G = \frac{-du_\theta}{dr} = A - \frac{B}{r^2} \tag{17}$$

The lowest shear rate at outer cylinder wall and the highest shear rate at inner cylinder wall at laminar flow are shown in **Table 3**. Due to the narrow gap of Taylor reactor, the shear rate difference between inner cylinder wall and outer cylinder wall is not significant. Therefore, shear rate in Taylor reactor can be seen as uniform, which can be reasonably expressed by the average shear rate. This feature is also one good performance of this reactor. However, when fluid becomes turbulent, shear rate is not only dominated by tangential velocity due to the enhanced radial and axial velocities induced by Taylor vortex. Also azimuthal velocity is different from the one at laminar state. For turbulent Taylor vortex flow, the local shear rate caused by turbulent eddies with Kolmogorov length scale can be estimated from turbulent energy dissipation rate, given by

$$G = \sqrt{\frac{\varepsilon}{\nu}} \tag{18}$$

The exact turbulent energy dissipation rate is difficult to determine but the mean turbulent energy dissipation can be estimated from the following equation for Taylor-Couette reactor [26].

$$\langle \varepsilon \rangle = \frac{P}{m_l} = \frac{\pi L r_i^4 \omega_i^3 f}{V} \tag{19}$$

In addition, the friction factor f is given as a function of δ_{eq}/r_i and the Reynolds number [39].

$$f = 0.80 \left(\frac{\delta_{eq}}{r_i} \right)^{0.35} Re^{-0.53} \tag{20}$$

Rotational speed (rpm)	Inner cylinder wall shear rate (s^{-1})	Outer cylinder wall shear rate (s^{-1})	Average shear rate (s^{-1})
25	12.17	9.56	10.87
50	24.35	19.11	21.73
100	48.70	38.23	43.46

Table 3.
Shear rate in laminar Taylor-Couette flow system.

The change of shear rate at different rotational speed for current study is shown in **Figure 3**. It can be seen from **Figure 3** that shear rate has a sharp increase when the flow pattern changes from laminar flow into turbulent flow.

5.1 Particle morphology at a low concentration of 0.1 mol/L

At a low concentration of 0.1 mol/L, the rotational speed was varied from 25 to 1000 rpm with the feeding rate being given ranging from 5 to 80 mL/min. There were three kinds of particle morphology observed from the SEM images in the same magnification as shown in **Figure 4**. The appearance of these particles looks like flake, transition mode and granule, respectively. **Figure 5(a)** shows the changes of morphology under the influence of the rotational speed and feeding rate. The obtaining of the transition mode particles was marked by connecting those data points that can be clearly distinguished from the other two states. It can be seen from the **Figure 5(a)** that the flake particles are mainly formed at low rotational speed while granule particles tend to be presented with the increase of rotational speed at low feeding rate. It is thus indicated from **Figure 5(a)** that the flake particles are expected to be generated when applying the high feeding rate (the top left region of line AB) and granule particles are formed with the high rotational speed (the bottom right region of line DE). However, it cannot be identified that whether or not there exists an intersection point with the development of line AB and line DE.

It has been suggested that beyond a transition Reynolds number at $Re_T = 1.3 \times 10^4$, the flow in the reactor becomes turbulent Taylor flow [40], this corresponding to the rotational speed of 313 rpm (with the aspect ratio Γ = 30.58) in current study. From **Figure 5(a)**, transition state appears at the rotational speed of 300 rpm, which is in a good agreement with the transition to turbulent Taylor flow as predicted.

5.1.1 Mechanism of granule particles formation

The formation of granule particles with high rotational speed can be explained in terms of the mixing efficiency and agglomerative bonding force. It can be seen from **Figure 5(a)** that the minimum value of rotational speed at which granule particles appear is 600 rpm, and the flow has already developed into fully turbulent Taylor flow. Also, the feeding rate is low enough to form granule particles at this point. As micro-mixing of the fresh feed and bulk solution has significant effects on the final products, the mixing process of two solutions can be explained by the micro-mixing

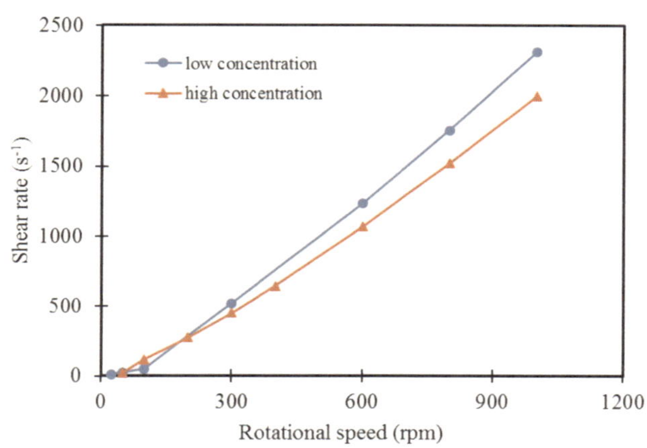

Figure 3.
Shear rate at different rotational speeds.

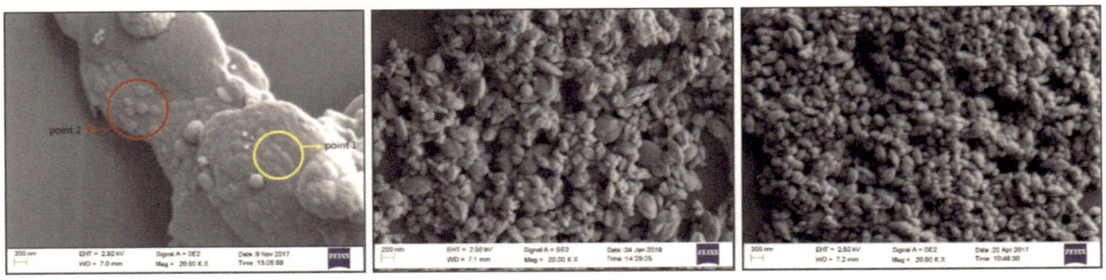

(a) (b) (c)

Figure 4.
Three representative morphologies of barium sulfate particles at low concentration. (a) ω = 25 rpm; Q = 20 mL/min, flake; (b) ω = 300 rpm; Q = 40 mL/min, transition; and (c) ω = 600 rpm; Q = 5 mL/min, granule.

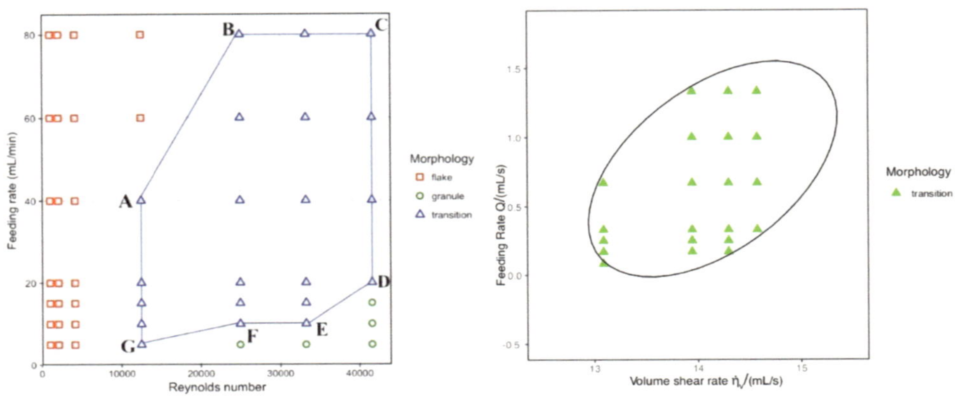

(a) (b)

Figure 5.
(a) Morphology distribution at various rotational speed (Reynolds number) and feeding rate at low concentration; (b) Morphology distribution for transition status where the transition boundary has been marked by curve fitting.

model. Many theoretical models have been proposed to describe the micro-mixing behavior in the reactors, such as diffusion model [41], deformation-diffusion model [42], eddy engulfment model [21], and incorporation model [43]. Among these models, the engulfment model proposed by Baldyga and Bourne [21] has been widely adopted. The engulfment rate is defined as

$$E = 0.058\left(\frac{\varepsilon}{\nu}\right)^{1/2} \tag{21}$$

Based on this model, two species are considered as two shrinking slabs [15]. In case of turbulent Taylor vortex flow, once the fresh reactant Na_2SO_4 has been added into the Taylor-Couette flow reactor, two reactant slabs are in contact and interact each other with the turbulent eddies being stretched due to turbulent vorticity and generating a local turbulent shear. The scale of eddies can be approximately considered as being proportional to Kolmogorov length scale, defined by

$$\lambda_{eddy} = C_{eddy}\left(\frac{\nu^3}{\varepsilon}\right)^{1/4} \tag{22}$$

Fluid velocity fluctuation increases with intensified turbulence. More small eddies generate and dissipate. It can be seen from Eq. (22) that the length scale is reduced with the increase of energy dissipation rate. The incorporation of fresh feed after finishing the processes of engulfment and deformation will finally become small fluid elements, smaller than Kolmogorov length scale. Accordingly, at high rotational speed, the contact area between fresh solution and bulk solution increases, leading to the acceleration of the molecular diffusion. The characteristic diffusion time can be estimated according to Baldyga and Bourne [44], given by

$$t_D = 2\left(\frac{\nu}{\varepsilon}\right)^{1/2} arcsinh(0.05S_C) \tag{23}$$

where *Sc* is Schmidt number. Based on this model, the characteristic time by engulfment on the course of mixing is roughly equal to E^{-1},

$$t_m = 12\left(\frac{\nu}{\varepsilon}\right)^{1/2} \tag{24}$$

In our experiment, this characteristic time (estimated to range from 0.711 to 0.0075 s based on different rotational speed) was about 17 times larger than diffusion time (estimated to range from 0.041 to 0.0004 s). Therefore, it is suggested that the engulfment may control the micro-mixing process which can be seen as the rate-determining step and t_m can be seen as the typical micro-mixing time.

The time constants are of importance in assessing the influence of mixing on reactive precipitation. It has been suggested by Baldyga and his co-workers [23] that when the mixing time, t_m is shorter than the induction time t_N, $t_m < t_N$, nucleation step cannot be affected by mixing intensity, as the local region has uniformly dispersed with reactants before nucleation. While when $t_m > t_N$, nucleation would occur in a non-uniform region. Under this condition, the effect of mixing on precipitation can be crucial. Since the $BaSO_4$ reactive precipitation is fast by nature, reactants contact quickly once the Na_2SO_4 has been added into the Taylor-Couette flow reactor, resulting in the generation of local supersaturation. Then, nuclei are formed in the duration of induction. The induction time was about 0.0014 s, which was smaller than the micro-mixing time 0.0075 s at 1000 rpm. Therefore, reactive precipitation can be controlled by mixing condition achieved by fluid shear. Because the induction period is independent with the micro-mixing process, the precipitation reaction takes place and the rest of fresh reactants is incorporated and diffused into the vortices at the same time. Therefore, the maximum nucleation rate attains at the first contact of two species due to the dilution of supersaturation. This in fact indicates that after a number of nuclei are formed, the growth rate decreases with the reduced supersaturation as there is little amount of reactant left for nuclei to grow further.

At the end of the primary process of crystallization, secondary process takes place, having a determining effect on particle qualities. Nuclei grow into grains then crystallites. This process is dependent on the mass transfer, while mass transfer is caused by interfacial concentration gradient. Convective mass transfer and diffusive mass transfer coexist in the system simultaneously. Here the Sherwood number, *Sh* derived by Armenante and Kirwan [28] as a function of Reynolds number and Schmidt number can be used to illustrate the interfacial mass transfer during nuclei growth.

$$Sh = 2 + 0.52Re^{0.52}Sc^{1/3} \tag{25}$$

As granule particles are formed at a relative high rotational speed, an increase in Sherwood number indicates an enhanced convection mass transfer from bulk solution to solid surface. It can be deduced that an intensified interfacial mass transfer can facilitate the formation of granular particles. Additionally, Peclet number, *Pe* is used here to determine the dominant effect on the interaction among particles, expressed as a function of Reynolds number and Schmidt number which is given by

$$P_e = R_e S_C = \frac{r_i \omega_i \delta_{eq}}{D} \tag{26}$$

where D is the diffusion coefficient, including the effects of Brownian motion and molecular diffusion. The range of Pe applicable in the present experiments was $9.7 \times 10^5 < Pe < 3.9 \times 10^7$, which was much greater than 1 ($Pe >> 1$). Larger Pe values indicate that the particle collision is dominated by the convective mass transfer due to Taylor turbulent eddy transport. Such convection effect can also be improved by the intensification of turbulent eddies, realized by increasing the rotational speed of the inner cylinder.

Due to the entrainment of these crystallites by the turbulent eddies, the self-assembled aggregation makes them closer, leading to the agglomeration of crystallites which further results in the formation of final particles. In Aljishi et al.'s study [6], they described crystallites as the primary grains and final particles as the secondary grains. The joint effect of micro-mixing and crystallization on the formation of the particles and its mechanism are illustrated in **Figure 6**. For powder-like particle preparation, agglomeration becomes an important growth mechanism due to a large number of nuclei created. The final particle morphology is a result of the relative rates of chemical reaction, nucleation, growth and agglomeration. During this process, the motion of two or more crystallites in suspension makes them closed and then aggregate. Furthermore, once the crystals have stayed together for a sufficient long time, the agglomerative bonds can be formed due to intergrowth [45]. An empirical formulation of the net agglomeration rate (including bond making and breaking) was proposed

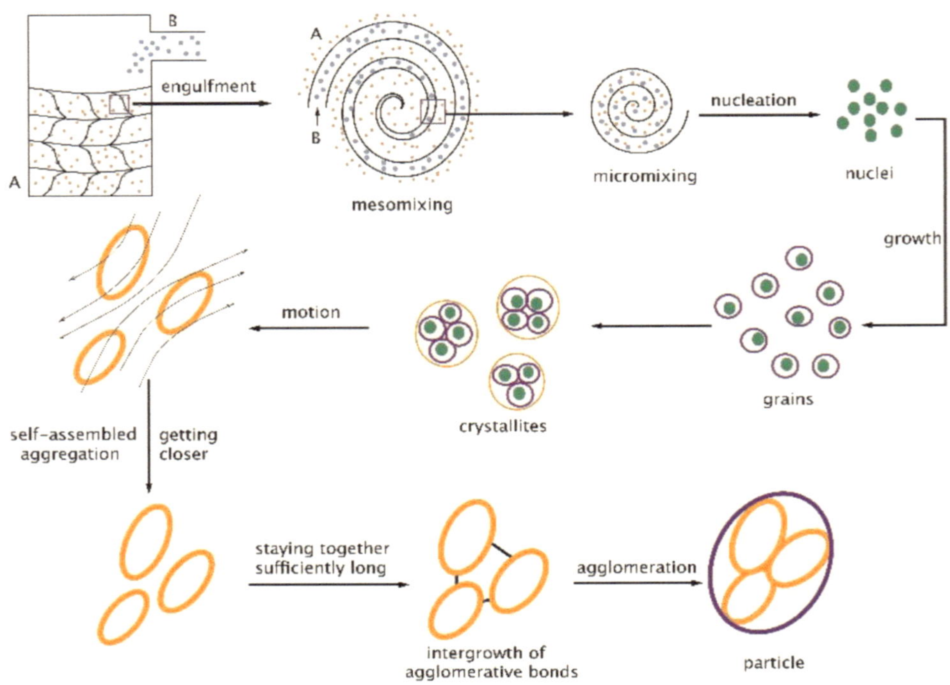

Figure 6.
Mechanism of particle formation in fluid.

$$K_{aggl} = \beta_{aggl}\varepsilon^{p}S^{q}f(l_m, l_n) \quad (27)$$

where β_{aggl} is agglomeration efficiency, l_m and l_n denote the sizes of the binary agglomerating particles. It can be seen clearly from this equation that the agglomeration rate is influenced by the turbulence dissipation rate and supersaturation. Therefore, for a given rotational speed, the average shear rate is a constant while along with the nucleation process based on our analysis before, the supersaturation is reduced. This can then result in the reduction of agglomeration rate. Accordingly, the particle size cannot increase continuously as the agglomeration rate reaches its limit at a given condition. This is consistent with the results obtained in the experiments. At the granule state, particles morphologies are similar and their size were kept at around 180 to 200 nm no matter the change of rotational speed. This result is consistent with the finding of Dehkordi and Vafaeimanesh [46]. They also obtained a constant final particle size. Moreover, especially at high rotational speed of 1000 rpm, application of the higher feeding rate can still promote the generation of granule particles because at this feeding rate, the mass dispersed is saturated while the reaction process remains the same, thus resulting in the particle size unchanged. This can explain why the particle morphologies and sizes are similar at granule state no matter how the feeding rate and rotational speed change.

In addition, at transition state, we also measured the particle size, and found that the particle size exhibits a reduction trend with increase in the rotational speed at a constant feeding rate as shown in **Figure 7**. Based on the Reynolds number calculated before, at the rotational speed of 300 rpm, turbulent Taylor vortex flow has been formed. Particle morphology shows a significant difference from the previous large blocks with irregular shape. The rotational speed of 300 rpm almost corresponds to the point where turbulent Taylor vortex flow began to form. Thus, the disturbance effect of feeding rate cannot be ignored. Fluid is sensitive to any small perturbation, as this perturbation will lead to big velocity fluctuation, then create small eddies. Aljishi et al. [6] observed in their experiment that with increase in axial flow intensity, the onset of flow pattern instability is significantly delayed. Also, it can be seen from **Figure 5(a)** that at 600 rpm, the beginning point of granule particles formation are observed at a low feeding rate, where the disturbance effect is weak. Therefore, it is suggested that the appearance of particle size reduction trend can be seen as an indicator for morphology transition.

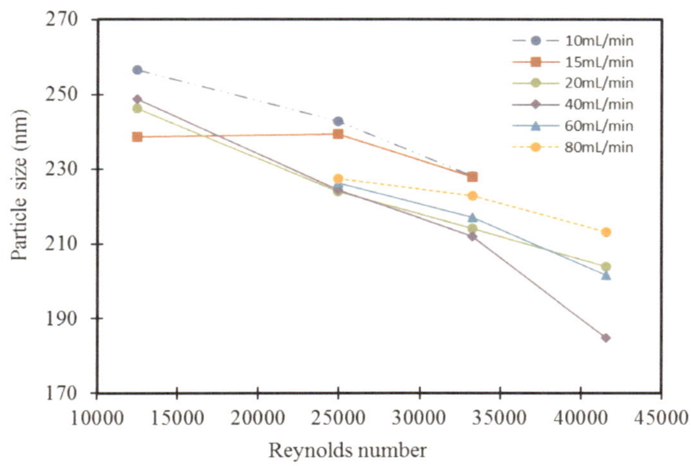

Figure 7.
Change of particle size with rotational speed (Reynolds number) at transition state at low concentration.

5.1.2 Mechanism of the formation of flake particles

It can be clearly seen from **Figure 4(a)** that the formation of big flake particles resulted from the stable agglomeration of smaller particles. The blurred boundary marked at point 1 and the small particles attached at point 2 indicate that the big flake particle is the product of small particles aggregation then agglomeration or level by level accumulation. This formation process is strongly influenced by the flow dynamics. In order to elaborate on this phenomena, **Figure 8** shows the morphology of the formed flake particles at various rotational speeds. **Figure 8(e)** also shows the transition mode particles for comparison.

Firstly, many small particles are formed and suspend in the mixture solution. When low rotational speed is applied, the entrainment on the fresh feed by eddies is not strong and the formed particles can disperse quite uniformly. Because at laminar state, velocity is dominated by azimuthal velocity. The effect of Taylor vortex is not significant. Therefore, mixing behavior is not strong. Accordingly, the contact area between the bulk solution and the fresh feed is not big enough instead more mass transfers to the formed small particle surface, leading to the local accumulation. The length scale of particles is smaller than minimum vortex scale. Fluid shear induced by the velocity gradient acted on particle agglomeration. Therefore, the small particles generated under this condition are still larger than those in high rotational speed. Also, weak shear decreases the engulfment of bulk solution and fresh feed, which means the exchange between the bulk solution entrained by eddies and fresh feed is poor at the inlet. This phenomenon becomes notable at a high feeding rate. For a particular small region, enhanced collision can promote the generation of agglomeration bonds. This kind of agglomeration is gradually reduced as shown from **Figure 8(b)–(e)**, consistent with the analysis above. It should be noted that as can be seen from **Figure 8(a)** where there is no influence of the turbulent eddies due to zero rotational speed of the inner cylinder, few formed small particles can be still observed. At this condition, the droplets of fresh Na_2SO_4 feed directly contact with bulk $BaCl_2$ solution from the inlet and are diffused into the bulk solution, the reactive product as agglomerated big flakes sinks to the

Figure 8.
Development of the flake particles with increase of the rotational speed. (a) ω = 0, flake; (b) ω = 25 rpm, flake; (c) ω = 50 rpm, flake; (d) ω = 100 rpm, flake; and (e) ω = 300 rpm, transition. Note: the magnification of image (a) is 2.00 k, and the others are 20.00 k.

bottom of reactor. This is likely caused by a local saturated diffusion in which many formed small particles agglomerate to form the big flakes.

Secondly, the effect of eddy shear on flake particles is different from granule particles. At high rotational speed, micro-mixing takes place at Kolmogorov length scale, which is larger than the size of granule particles. Therefore, larger turbulent eddies only affect the mixing process of two species, indicating that the particles passively follow the vortex motion. Here we use Stokes number, *St* to describe the trapping of particles by an eddy [47].

$$\mathrm{St} = \frac{\rho_p d^2 u_t}{18\mu l} \tag{28}$$

where u_t is the terminal velocity of particles [48], and l is the suitable length scale for the vortex. For small eddies in Taylor reactor, l can be chosen as Kolmogorov length scale for the calculation of St. At high rotational speed for the formation of granule particles with the diameter of 200 nm, *St* is approximately 1.1×10^{-9}, which is much smaller than 1.0. As indicated by Crow et al. [42], $St << 1.0$ represents the case that particles are well following the eddy. However, for low rotational speed, Kolmogorov length scale increases, and the scale of formed flake particles is in the order of eddies. Vortex and flake particles are all at micron scales. Therefore, eddy shear strongly interacts with the flake particles. If shear force is strong enough, the size of flake particles will be smaller than Kolmogorov length scale as the particle surface will be ripped off. This is consistent with the observation from the semi-batch precipitation experiment of Baldyga et al. [15] in which they suggested that rotational speed had no influence on particle size at a relatively weak turbulence. Barresi et al. [12] also agreed with this argument from their variation coefficient analysis.

Based the analysis above, both rotational speed and feeding rate have effects on final particle morphology. Also, the transition region in **Figure 5(a)** indicates an ellipse-like shape when the both parameters are combined. In order to obtain quantitative information about particle morphology, a fitting equation is proposed involving shear rate (rotational speed) and feeding rate, shown in **Figure 5(b)**.

$$0.33\dot{\eta}_V^2 - 0.51\dot{\eta}_V Q + 0.79Q^2 - 9.06\dot{\eta}_V + 6.02Q + 61.41 = 0 \tag{29}$$

The relationship is expressed in a semi-natural log coordinate. In order to unify the dimension, shear rate here refers to the entire volume shear rate, $\dot{\eta}_V$ for the Taylor reactor, defined as.

$$\dot{\eta}_V = GV \tag{30}$$

Eq. (28) can be seen as an indicator for particle morphology. Inside the ellipse, transition mode takes place, while in the top left region of the ellipse, flake particles can be created, and in the bottom right region, granule particles are formed.

5.2 Particle morphology at a high concentration of 1.0 mol/L

In order to elaborate on this transition phenomenon, experiments at a higher concentration of 1.0 mol/L were also carried on. The rotational speed was varied from 50 to 1000 rpm, and the feeding rate was changed from 5 to 60 mL/min. Three similar morphologies as at low concentration were observed as shown in **Figure 9**.

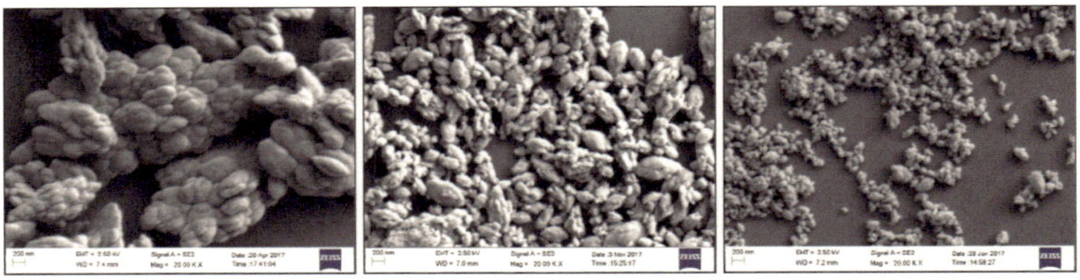

(a) (b) (c)

Figure 9.
Three representative morphologies of barium sulfate particles at high concentration. (a) ω = 50 rpm; Q = 5 mL/min, flake; (b) ω = 100 rpm; Q = 5 mL/min, transition; and (c) ω = 800 rpm; Q = 10 mL/min, granule.

Also, a transition region was identified to distinguish the other two states as shown in **Figure 10**. Compared with that shown in **Figure 5(a)** at 0.1 mol/L, the transition region is exhibited to be narrower while the beginning points and ending points fall into the range of relatively lower rotational speeds. Particle size at transition state were measured, also illustrating a reduction trend as shown in **Figure 11**. Thus, the size reduction can be considered as an indicator for morphology transition.

At low concentration, the change of kinematic viscosity means the change of Reynolds number even at the same rotational speed. The kinematic viscosity at 1.0 mol/L is about 1.85 times as high as 0.1 mol/L. According to the radius ratio of η = 0.80, the critical Reynolds number to generate Taylor vortex flow is 97.33 [34]. The generation of turbulent Taylor vortex flow is at a Reynolds number ratio of $R = Re/Re_c \sim 35$ as indicated by Bubakova et al. [29], corresponding to the rotational speed of 89 rpm. Therefore, turbulent Taylor vortex flow occurs at the rotational speed of 89 rpm, roughly corresponding to the beginning of transition state at 100 rpm in this case. Also, fully turbulent flow is formed at the rotational speed of 581 rpm, roughly corresponding to the ending of transition state with the rotational speed of 600 rpm. We thus deduce that the morphology transition is correlated with the flow patterns in the reactor. Based on engulfment model, a good mixing of

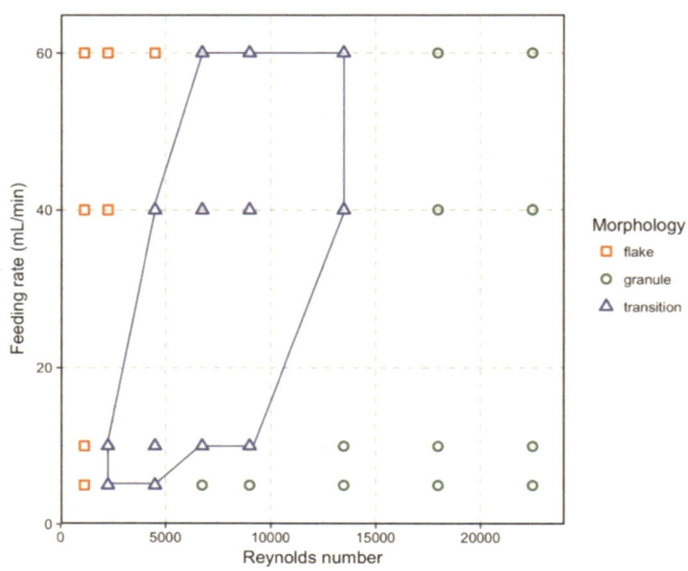

Figure 10.
Morphology distribution at various rotational speed (Reynolds number) and feeding rate at high concentration.

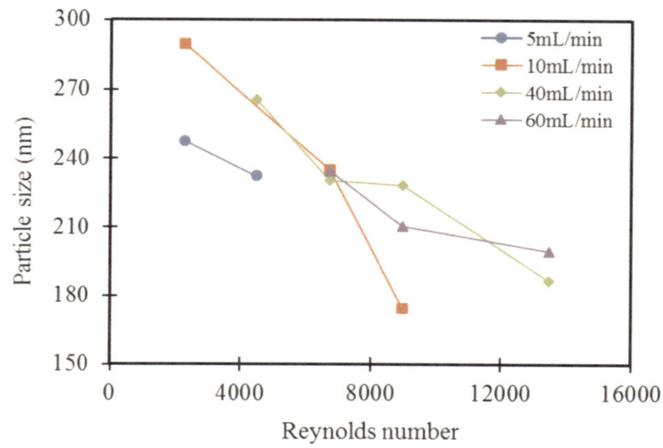

Figure 11.
Change of particle size with rotational speed (Reynolds number) at transition state at high concentration.

two species needs high energy input. As mentioned earlier, formation of flake particles is the result of the poor dispersion of reactants. When the transition was observed, the micro-mixing was improved. In the present work, the transition state was observed at least under the condition of turbulent flow.

5.3 Influence of the low concentration and high concentration on particle formation

It was noticed that the onset of the transition state is different for different concentrations. At low concentration, transition state was observed at turbulent flow regime at 300 rpm while at high concentration, such transition appears at turbulent Taylor vortex flow regime at 100 rpm. Also, the transition range becomes smaller at high concentration. To keep the Reynolds number unchanged, the increase in kinematic viscosity means that the rotational speed has to be increased to maintain the same fluid pattern. This implies that the transition state should be observed at a higher rotational speed than that in low concentration. Also, based on the mechanism of various morphologies formation at low concentration, the onset of turbulent flow is essential for the creation of transition particles. However, at high concentration, transition state appeared at a much lower rotational speed. The final granule particle size at high concentration was approximately measured as 130–150 nm, which is smaller than that at low concentration.

The above observed results should also result from the increase in local supersaturation, which was the only varying parameter in the comparison experiments. Since the nucleation and growth are directly dependent on supersaturation, the effect of supersaturation is crucial on morphology transition. As can be seen from Eqs. (10) and (11), the order of magnitude of nucleation rate is much higher than growth rate, indicating that nucleation is more sensitive to supersaturation than growth. Many studies have been suggested that an increase in supersaturation may result in the nucleation process to be dominant [12]. In the experiments of Barresi et al. [12] and Marchisio et al. [13], they all observed tabular particles to form at low concentration and dendritic particles at high concentration. Baldyga et al. [15] suggested that the formation of tabular shape is the consequence of surface nucleation, leading to the overgrowth of crystal surface. While with improvement of flow environment (increasing flow rate in jet) round-shaped particles can form [13]. In our experiment, granule particles were observed with the increase of the rotational speed (likely corresponding a strong local turbulent shear due to the turbulent eddies). Since with the increase in supersaturation, the mass brought into the system is more than that at low supersaturation for the same small droplets. In such environment, the available reactant molecules mainly function to form nuclei. Thus, the nucleation significantly increases. It is thus concluded that a higher nucleation rate facilitates the creation of a large number of nuclei which is beneficial to the formation of small particles. On the other hand, as mentioned in the part 5.1.1, an increased interfacial concentration gradient can facilitate the formation of granular morphology. Therefore, in general, a high supersaturation results in the formation of small granule particles. It should be noticed that agglomeration is an important part on the course of nucleation. It is usually significant at low mixing rate [46]. At low rotational speed, agglomerative bonding force is strengthened due to weak shear and large vortex scale, more primary crystallites agglomerate to form large flake particles as discussed earlier. It can thus be inferred that the supersaturation can be seen as a primary factor in determining particle morphology while the effect of agglomeration is the second place but it cannot be ignored at low rotational speed. Furthermore, from the analysis of particle formation mechanism mentioned above, the creation of agglomerative bonds requires a close distance and sufficient long time. With the increase of rotational speed, fluid motion becomes more significant. The formation of agglomerative bonds was reduced, followed with the number decrease of crystallites to agglomerate. Therefore, it is suggested that the appearance of particle size reduction trend can be seen as an indicator for morphology transition.

Based on the analysis and discussion above, it can be suggested that when using Taylor-Couette flow reactor for synthesis of the particles, mixing affects the whole process especially for the mixing of individual reactant solution through the engulfment and the subsequent chemical reaction. Supersaturation has the effect on crystallization process when reactants are in contact but this effect may surpass the flow environment. It is conjectured that agglomeration can be seen as a consequence of flow dynamics and supersaturation when working on crystallites. Last but not the least, interfacial phenomenon between liquid and solid particles exists throughout the whole precipitation process. It includes mainly two aspects. One is from the interfacial concentration distribution, which determines the mass transfer process from liquid to solid, thus nuclei growth process, and the other is the surface treatment of fluid shear on final agglomerates.

6. Conclusions

In the present work, a Taylor-Couette reactor was employed for the purpose of controllable production of barium sulfate particles with various morphologies. The morphologies obtained include flake particles, transition particles and granule particles. During the reactive precipitation process, three main control parameters were assessed. The presence of a particular morphology results from the combined effect of these parameters. For different solution conditions, the influence of fluid dynamic environment inside the Taylor-Couette reactor on particle morphology may change. In general, morphology transition is suggested to be related to the change of flow pattern that the turbulent eddies strongly interact with the particle crystals. Considering the joint effect of fluid dynamics and supersaturation, it can be deduced that micro-mixing caused by fluid shear plays a primary role in species dispersion, which determines the formation of various morphologies, while the supersaturation controls the crystallization process, which determines the onset of transition state. Both fluid shear and supersaturation can finally affect the interfacial concentration distribution, thus particle growth. Then particle agglomeration can be controlled by interfacial shear force. It is thus concluded that interfacial phenomenon plays significant roles in the transition of particle morphologies during precipitation. More specifically, the following conclusions have been drawn from the current study:

1. At low supersaturation, the morphology transition mode is observed at fully developed turbulent Taylor flow, and it appears at turbulent Taylor vortex flow at high concentration. It is indicated that turbulent shear creating isotropic small eddies can assist controllable synthesis of particles, which results in the generation of relatively regular morphology. This is due to the enhanced mixing intensity and interfacial mass transfer caused by shear at turbulent flow. However, anisotropic vortex shear in laminar state has the potential to create irregular particle morphology. On the other hand, agglomeration as a part of crystallization, can be enhanced by both low mixing intensity and high supersaturation, leading to the formation of flake particles.

2. The decrease in particle size with the increase of rotational speed at transition state results from the disturbance effect of the feeding rate. Fluid is sensitive to any small perturbation especially at high rotational speed, then more small isotropic eddies will be created.

3. Compared the transition results at both low and high concentrations, supersaturation will lead itself to the formation of smaller particles, and the transition at lower rotational speed. Although the use of high supersaturation can increase the agglomeration for a given rotational speed, it facilitates the formation of a large number of nuclei but leaves little mass for further particle growth. A high interfacial concentration also enhances the mass transfer between fluid and solid particles. Thus, it can be inferred that the supersaturation has a stronger effect than the agglomeration in determining the particle morphology.

Nomenclature

a	ion activity, mol/kg
A_i	cross sectional area of the inner cylinder, m
B	growth rate, m/s
c	molar concentration, mol/L
Δc	concentration difference, mol/L
d	diameter of particle, m
D	molecular diffusion coefficient, m^2/s
E	engulfment rate, s^{-1}
f	friction factor
G	shear rate, s^{-1}
I	ionic strength, mol/kg
J	nucleation rate, m/s
k_{aggl}	agglomeration rate
K_g	total growth coefficient, m/s
K_{sp}	thermodynamic solubility product, mol^2/L^2
l	length scale, m
l_m	size of agglomerating particle m, m
l_n	size of agglomerating particle n, m
L	reactor length, m

n	kinetic order of nucleation
Pe	Peclet number
Q	feeding rate, mL/min
r_i	equivalent radius of the lobed inner cylinder, m
r_o	radius of the lobed outer cylinder, m
R	gas constant, J/(mol·K)
Re	Reynolds number
R_i	perimeter of the inner cylinder, m
S	saturation ratio
Sa	supersaturation ratio
Sc	Schmidt number
Sh	Sherwood number
St	Stokes number
t_D	diffusion time, s
t_N	induction time, s
t_m	micro-mixing time, s
T	absolute temperature, K
u_t	terminal velocity of particle, m/s
u_θ	azimuthal velocity, m/s
V	volume of reactor, m^3
z	charge number

Greek letters

β_{aggl}	agglomeration efficiency
$\gamma_\pm$	activity coefficient
δ_{eq}	equivalent gap, m
Γ	aspect ratio
ε	turbulent energy dissipation rate, m^2/s^2
ζ	angular velocity ratio
η	radius ratio
$\dot{\eta}_V$	volume shear rate, mL/s
λ_{eddy}	eddy length scale, m
μ	viscosity of the fluid, kg/(m·s)
ρ_d	density of particle, kg/m^3
σ	supersaturation
υ	kinematic viscosity of the fluid, m^2/s
φ	molar chemical potential, J/mol
ω_i	rotational speed of the lobed inner cylinder, rad/s

Acknowledgements

This work was financially supported by National Natural Science Foundation of China (NSFC) through the grants (Nos. 21576141, 21606259). This work was carried out at the International Doctoral Innovation Centre (IDIC), UNNC. Lu Liu would also like to acknowledge the support through the Ph.D. scholarship of the International Doctoral Innovation Centre (IDIC) of University of Nottingham Ningbo China.

Author details

Lu Liu[1,2], Guang Li[1,2], Xiaogang Yang[1,2]*, Xiani Huang[1,2] and Chenyang Xue[1,2]

1 International Doctoral Innovation Centre, University of Nottingham Ningbo China, Ningbo, P.R. China

2 Department of Mechanical, Materials and Manufacturing Engineering, University of Nottingham Ningbo China, Ningbo, P.R. China

*Address all correspondence to: xiaogang.yang@nottingham.edu.cn

References

[1] Tai CY, Chen FB. Polymorphism of CaCO3, precipitated in a constant-composition environment. AICHE Journal. 1998;**44**(8):1790-1798

[2] Tai CY, Chen PC. Nucleation, agglomeration and crystal morphology of calcium carbonate. AICHE Journal. 1995;**41**(1):68-77

[3] Jung WM, Kang SH, Kim WS, Choi CK. Particle morphology of calcium carbonate precipitated by gas–liquid reaction in a Couette–Taylor reactor. Chemical Engineering Science. 2000; **55**(4):733-747

[4] Akyol E, Cedimagar MA. Size and morphology controlled synthesis of barium sulfate. Crystal Research and Technology. 2016;**51**(6):393-399

[5] Judat B, Kind M. Morphology and internal structure of barium sulfate—derivation of a new growth mechanism. Journal of Colloid and Interface Science. 2004;**269**(2):341-353

[6] Aljishi MF, Ruo AC, Park JH, Nasser B, Kim WS, Joo YL. Effect of flow structure at the onset of instability on barium sulfate precipitation in Taylor–Couette crystallizers. Journal of Crystal Growth. 2013;**373**:20-31

[7] Farahani HB, Shahrokhi M, Dehkordi AM. Experimental investigation and process intensification of barium sulfate nanoparticles synthesis via a new double coaxial spinning disks reactor. Chemical Engineering and Processing: Process Intensification. 2017;**115**:11-22

[8] Wu G, Zhou H, Zhu S. Precipitation of barium sulfate nanoparticles via impinging streams. Materials Letters. 2007;**61**(1):168-170

[9] Nielsen AE. Nucleation in barium sulfate precipitation. Acta Chemica Scandinavica. 1957;**11**:1512-1515

[10] Nielsen AE. The kinetics of crystal growth in barium sulfate precipitation. Acta Chemica Scandinavica. 1958;**12**(5): 951-958

[11] Nielsen AE. The kinetics of crystal growth in barium sulfate precipitation. Acta Chemica Scandinavica. 1959;**13**: 1680-1686

[12] Barresi AA, Marchisio D, Baldi G. On the role of micro-and mesomixing in a continuous Couette-type precipitator. Chemical Engineering Science. 1999;**54** (13–14):2339-2349

[13] Marchisio DL, Barresi AA, Garbero M. Nucleation, growth, and agglomeration in barium sulfate turbulent precipitation. AICHE Journal. 2002;**48**(9):2039-2050

[14] Li J, Liu D, Jiang H, Wang J, Jing X, Chen R, et al. Effects of polyacrylic acid additive on barium sulfate particle morphology. Materials Chemistry and Physics. 2016;**175**:180-187

[15] Baldyga J, Podgórska W, Pohorecki R. Mixing-precipitation model with application to double feed semibatch precipitation. Chemical Engineering Science. 1995;**50**(8):1281-1300

[16] Baldyga J, Orciuch W. Closure problem for precipitation. Chemical Engineering Research and Design. 1997; **75**(2):160-170

[17] Phillips R, Rohani S, Baldyga J. Micromixing in a single-feed semi-batch precipitation process. AICHE Journal. 1999;**45**(1):82-92

[18] Pagliolico S, Marchisio D, Barresi AA. Influence of operating conditions on $BaSO_4$ crystal size and morphology in a continuous Couette precipitator. Journal of Thermal Analysis and Calorimetry. 1999;**56**(3):1423-1433

[19] Thai DK, Mayra QP, Kim WS. Agglomeration of Ni-rich hydroxide crystals in Taylor vortex flow. Powder Technology. 2015;**274**:5-13

[20] Mayra QP, Kim WS. Agglomeration of Ni-rich hydroxide in reaction crystallization: Effect of Taylor vortex dimension and intensity. Crystal Growth & Design. 2015;**15**(4):1726-1734

[21] Baldyga J, Bourne JR. A fluid mechanical approach to turbulent mixing and chemical reaction Part III computational and experimental results for the new micromixing model. Chemical Engineering Communications. 1984;**28**(4–6):259-281

[22] Vicum L, Mazzotti M, Baldyga J. Applying a thermodynamic model to the non-stoichiometric precipitation of barium sulfate. Chemical Engineering & Technology. 2003;**26**(3):325-333

[23] Baldyga J. Mixing and fluid dynamics effects in particle precipitation processes. Kona Powder and Particle Journal. 2016;**33**:127-149

[24] Li G, Yang X, Ye H. CFD simulation of shear flow and mixing in a Taylor–Couette reactor with variable cross-section inner cylinders. Powder Technology. 2015;**280**:53-66

[25] Brian PLT, Hales HB, Sherwood TK. Transport of heat and mass between liquids and spherical particles in an agitated tank. AICHE Journal. 1969;**15**(5):727-733

[26] Kuboi R, Komasawa I, Otake T. Behavior of dispersed particles in turbulent liquid flow. Journal of Chemical Engineering of Japan. 1972;**5**(4):349-355

[27] Garside J, Jančić SJ. Growth and dissolution of potash alum crystals in the subsieve size range. AICHE Journal. 1976;**22**(5):887-894

[28] Armenante PM, Kirwan DJ. Mass transfer to microparticles in agitated systems. Chemical Engineering Science. 1989;**44**(12):2781-2796

[29] Bubakova P, Pivokonsky M, Filip P. Effect of shear rate on aggregate size and structure in the process of aggregation and at steady state. Powder Technology. 2013;**235**:540-549

[30] Debye P, Hückel E. De la theorie des electrolytes. I. abaissement du point de congelation et phenomenes associes. Physikalische Zeitschrift. 1923;**24**(9):185-206

[31] Dirksen JA, Ring TA. Fundamentals of crystallization: Kinetic effects on particle size distributions and morphology. Chemical Engineering Science. 1991;**46**(10):2389-2427

[32] Carosso PA, Pelizzetti E. A stopped-flow technique in fast precipitation kinetics—the case of barium sulphate. Journal of Crystal Growth. 1984;**68**(2):532-536

[33] Wereley ST, Lueptow RM. Spatio-temporal character of non-wavy and wavy Taylor–Couette flow. Journal of Fluid Mechanics. 1998;**364**:59-80

[34] DiPrima RC, Eagles PM, Ng BS. The effect of radius ratio on the stability of Couette flow and Taylor vortex flow. Physics of Fluids. 1984;**27**(10):2403-2411

[35] Desmet G, Verelst H, Baron GV. Local and global dispersion effects in Couette-Taylor flow—I. Description and modeling of the dispersion effects. Chemical Engineering Science. 1996;**51**(8):1287-1298

[36] Takeda Y. Quasi-periodic state and transition to turbulence in a rotating Couette system. Journal of Fluid Mechanics. 1999;**389**:81-99

[37] Haut B, Amor HB, Coulon L, Jacquet A, Halloin V. Hydrodynamics and mass

transfer in a Couette–Taylor bioreactor for the culture of animal cells. Chemical Engineering Science. 2003;**58**(3–6): 777-784

[38] Soos M, Wu H, Morbidelli M. Taylor-Couette unit with a lobed inner cylinder cross section. AICHE Journal. 2007;**53**(5):1109-1120

[39] Atsumi K, Makino T, Kato K, Murase T, Iritani E, Chidphong P, et al. Frictional resistance of grooved rotor in cylindrical dynamic filter chamber without permeation or throughflow. Kagaku Kogaku Ronbunshu. 1988;**14**(1): 14-19

[40] Lewis GS, Swinney HL. Velocity structure functions, scaling, and transitions in high-Reynolds-number Couette-Taylor flow. Physical Review E. 1999;**59**(5):5457

[41] Mao KW, Toor HL. A diffusion model for reactions with turbulent mixing. AICHE Journal. 1970;**16**(1): 49-52

[42] Ottino JM, Ranz WE, Macosko CW. A lamellar model for analysis of liquid-liquid mixing. Chemical Engineering Science. 1979;**34**(6):877-890

[43] Fournier MC, Falk L, Villermaux J. A new parallel competing reaction system for assessing micromixing efficiency—experimental approach. Chemical Engineering Science. 1996; **51**(22):5053-5064

[44] Baldyga J, Bourne JR. Principles of micromixing. Encyclopedia of fluid mechanics. Vol. 11986. Houston, TX: Gulf Publishing Co. pp. 147-201

[45] Brunsteiner M, Jones AG, Pratola F, Price SL, Simons SJ. Toward a molecular understanding of crystal agglomeration. Crystal Growth & Design. 2005;**5**(1): 3-16

[46] Dehkordi AM, Vafaeimanesh A. Synthesis of barium sulfate nanoparticles using a spinning disk reactor: Effects of supersaturation, disk rotation speed, free ion ratio, and disk diameter. Industrial & Engineering Chemistry Research. 2009;**48**(16): 7574-7580

[47] Crowe CT, Troutt TR, Chung JN. Numerical models for two-phase turbulent flows. Annual Review of Fluid Mechanics. 1996;**28**(1):11-43

[48] Haider A, Levenspiel O. Drag coefficient and terminal velocity of spherical and nonspherical particles. Powder Technology. 1989;**58**(1):63-70

Chapter 7

Dependence of Surface Tension and Viscosity on Temperature in Multicomponent Alloys

Ivaldo Leão Ferreira, José Adilson de Castro and Amauri Garcia

Abstract

Viscosity modeling for pure metals and alloys is widely studied, and many solutions for dependence of viscosity on temperature can be found in the literature for pure metals and alloys. Many of these depend on experimental data for pre-exponential and exponential coefficients. Two key models include: (i) Kaptay model for pure metals, which is completely independent of experimental data and depends only on general constants A and B for a large set of pure metals with few exceptions and (ii) Kaptay viscosity model for liquid alloys derived on the same principles, a temperature-dependent viscosity only as a function of thermophysical properties of the alloy components. In the case of surface tension, the main available models are divided into four groups: Butler formulation-based models, density-functional models, semi-empirical models, and thermodynamic geometric models. Considering the absence of adequate models for surface tension, in this work, two equations relating surface tension and viscosity for pure metals are analyzed as a function of temperature. Regarding the Egry surface tension-viscosity relation for pure metals, a new relation equation for multicomponent alloys is proposed. By applying the proposed equation, the surface tension is calculated and plotted as a function of temperature for ternary and quaternary aluminum alloys.

Keywords: viscosity, surface tension, thermophysical properties, multicomponent alloys, surface tension-viscosity correlation equation

1. Introduction

Surface tension is a phenomenon that occurs whenever a liquid is in contact with other liquids or even gases; then, an interface is established and it acts like a stretched elastic membrane. A surface is called wet if the contact angle is less than 90° and nonwet otherwise. A substrate, like dust or pollution, can contribute energetically to the membrane, decreasing its contact angle. Another feature is the magnitude of the surface tension σ $(N \cdot m^{-1})$. An important effect is the creation of curved meniscus, leading to capillary rise or depression. Viscosity, on the other hand, is a measure of the certain fluid's resistance to flow due to its internal friction. A high viscosity fluid has a trend to resist its motion, such as engine oil, while a fluid with low viscosity flows easily, such as water. Viscosity is a function of fluid's shear stress and its velocity gradient, and its magnitude η is expressed in $(Pa \cdot s)$. Both viscosity and surface tension are thermodynamic properties of a fluid, and, consequently, can be derived by means of thermodynamic relations.

Many solutions can be found in the literature concerning the viscosity dependence on temperature for pure metals [1–7] and for alloys [8–15]. Budai et al. have reviewed the existing models used to predict dependence of viscosity on temperature of alloys, for cases where the viscosities of pure components are already known [10, 15–18] and those that are independent of experimental data [19]. Budai et al. extended the Kaptay unified equation [10] for the viscosity of alloys, which has been named BBK model [20]. The BKK model was shown to fail in the prediction of viscosity for alloy systems with components that melt congruently [21].

Solutions for surface tension as a function of temperature are generally based on: Butler formulation [22–25]; statistical thermodynamics surface density-functional theory [26–28]; semi-empirical thermodynamic model [29]; and thermodynamic models [30, 31]. All these models are normally specific for certain binary or ternary alloy systems, or they are general but considerably difficult to apply.

In 1992, Egry [32] derived a relation between surface tension and viscosity deduced from statistical mechanics for the melting temperature, based upon the expressions of Fowler [26] for surface tension and Born and Green [33] for viscosity. Both expressions are expressed as integrals over the product of interatomic forces and the pair distribution functions. The author extended this relation to a finite temperature range by using data available in the literature [34].

In this work, by using a straightforward solution for viscosity for molten pure metals [7] and alloys [10], a comparison between numerical simulations and experimental data for the surface tension and viscosity of pure liquid metals and liquid alloys is provided, in order to validate Egry's relation for pure molten metals [34]. An extension of this relation is derived for multicomponent alloys. The surface tension is calculated and plotted against temperature for ternary and quaternary aluminum alloys.

2. Modeling

The modeling section is divided into models dealing with the viscosity of pure metals and multicomponent alloys, and with the surface tension-viscosity relation equations for pure liquid metals and alloys.

2.1 Model for the viscosity of pure liquid metals

Kaptay [7] derived a unified equation for the viscosity of pure liquid metals as a function of temperature, which encompasses the activation energy and the free volume concept. Based on the Andrade's equation [35] as a starting approach, the activation energy concept has been incorporated. By combining again with Andrade's formulation with free volume concept, an equation for the dependence of viscosity of pure metals on temperature has been derived. As this equation obeys both concepts, the authors named it as a unified equation for the viscosity of pure metals. The derived equation for the viscosity of pure metals as a function of temperature is given by,

$$\eta_i = A \frac{M_i^{1/2} \cdot T^{1/2}}{V_i^{2/3}} \cdot \exp\left(B \cdot \frac{T_{m,i}}{T}\right) \tag{1}$$

where η_i is the viscosity of the liquid metal i, A and B are temperature-independent semi-empirical parameters being approximately identical for all liquid metals, M_i ($kg \cdot mol^{-1}$) is the atomic weight of the metal, V_i is the molar volume ($m^3 \cdot mol^{-1}$), $T_{m,i}$ is the melting temperature of the pure liquid metal i (K), and T is the temperature above the melting point (K).

2.2 Model for the viscosity of multicomponent alloys

An Arrhenius-type viscosity equation can be extended to deal with viscosity of multicomponent alloy, by applying Redlich-Kister polynomial to excess viscosity,

$$\Delta\eta^E = \sum_i \sum_{j>i} x_i \cdot x_j \sum_{k=0}^{m} A_{i,j}^k (x_i - x_j)^k \tag{2}$$

where x_i and x_j are the molar contents of the solute compounds "i" and "j," respectively. $A_{i,j}^k$ are the polynomial parameters related to a binary "ij" system. The ideal viscosity term can be expressed as

$$\eta^{ideal}(T, x_i, x_j) = \sum_{i=1}^{3} x_i \cdot \eta_0^i \exp\left(\frac{E_A^i}{\overline{R} \cdot T}\right) \tag{3}$$

where η_0 is the pre-exponential factor independent of the temperature, interpreted as an asymptotic viscosity at very high temperature, E_A^i is the activation energy of viscous flow of the component i. By combining Eqs. (2) and (3), we obtain,

$$\eta(T, x_i, x_j) = \eta^{ideal}(T, x_i, x_j) + \Delta\eta^E \tag{4}$$

then, we have,

$$\eta(T, x_i, x_j, \cdots, x_n) = \sum_{i=1}^{3} x_i \cdot \eta_0^i \exp\left(\frac{E_A^i}{\overline{R} \cdot T}\right) + \sum_i \sum_{j>i} x_i \cdot x_j \sum_{k=0}^{m} A_{i,j}^k (x_i - x_j)^k \tag{5}$$

where $A_{i,j}^k$ are parameters for the viscosity of binary systems, according to Zhang et al. [36].

Kaptay [10], based on the Seetharaman-Du Sichen equation, regarding the theoretical relationship between the cohesion energy of the alloy and the activation of viscous flow, proposed the following equation:

$$\eta(T, x_i, x_j, \cdots, x_n) = \frac{h \cdot N_{Av}}{\sum_i x_i \cdot V_i + \Delta V^E} \cdot \exp\left[\frac{\sum_i x_i \cdot \Delta G_i^* - \alpha \cdot \Delta H}{\overline{R} \cdot T}\right] \tag{6}$$

where h is the Planck constant ($6.626 \times 10^{-34}\ J.s$), N_{Av} is the Avogadro number (mol^{-1}), ΔG_i^* is the Gibbs energy of activation of the viscous flow ($J \cdot mol^{-1}$), defined as

$$\Delta G_i^* = \overline{R} \cdot T \cdot \ln\left[\frac{\eta_i \cdot M_i}{h \cdot N_{Av} \cdot \rho_i}\right] \tag{7}$$

ΔV^E is the excess molar volume on alloy formation ($m^3 \cdot mol^{-1}$) and α is a ratio of a two properties ratios related to the melting temperature. The first ratio is between the measured activation energy and the melting point of pure liquid metals ($38.4 \pm 2.7\ J \cdot mol^{-1} \cdot K^{-1}$), and the second ratio is the cohesion energy of pure liquid metals and the melting points ($248 \pm 17\ J \cdot mol^{-1} \cdot K^{-1}$), providing $\alpha \cong 0.155 \pm 0.015$ [10].

2.3 Relation between surface tension and viscosity for pure liquid metals

Egry [32] derived a relation between surface tension and viscosity deduced from statistical mechanics for a finite temperature range, based upon the expressions of Fowler [26] for surface tension (σ) and Born and Green [33] for viscosity. Both expressions are expressed as integrals over the product of interatomic forces and the pair distribution functions. The Fowler expression is

$$\sigma = \frac{\pi\, n^2}{8} \int_0^\infty dR \cdot R^4 \, \frac{d\varphi(R)}{dR} \cdot g(R) \tag{8}$$

where n is the particle number density, φ is the pair potential, and $g(R)$ is the correlation function. All these functions depend on the temperature.

In a very similar way, Born and Green [33] derived an expression for the viscosity (η) of a fluid using a kinetic theory, which can be expressed as

$$\eta = \sqrt{\frac{m}{k\,T}} \cdot \frac{2\pi\, n^2}{15} \int_0^\infty dR \cdot R^4 \, \frac{d\varphi(R)}{dR} \cdot g(R) \tag{9}$$

where m is the atomic mass, k is the Boltzmann's constant, and T is the absolute temperature. As the integral terms of both integrals cancel each other, Egry [34] deduced the following relationship between density and viscosity for pure metals for a finite temperature above the melting point:

$$\frac{\sigma_i}{\eta_i} = \frac{15}{16} \sqrt{\frac{k\,T}{m}} \tag{10}$$

where σ_i is the surface tension in $(N.m^{-1})$, and η_i is the viscosity in $(Pa.s)$ of a pure liquid metal i.

In 2005, Kaptay [7] derived a surface tension-viscosity relation, as

$$\frac{\sigma_i}{\eta_i} = \frac{0.182 \cdot \left(211 + C_{p,i}\right) \cdot T_{m,i} - \left(2 + 0.182 \cdot C_{p,i}\right) \cdot T}{1.61 \cdot M_i^{1/2} \cdot \exp\left(2.34 \cdot T_{m,i}/T\right)} \tag{11}$$

where $C_{p,i}$ is the heat capacity of the pure liquid metal i in $\left(J.mol^{-1}K^{-1}\right)$.

3. A surface tension and viscosity relation model for multicomponent alloys

Based on the formula originally proposed by Egry [34], a new formula is derived for the relation between surface tension and viscosity for multicomponent alloys, that is,

$$\frac{\sigma_a}{\eta_a} = \frac{15}{16} \sqrt{N_{av}\, k\, T} \cdot \frac{1}{\sqrt{\sum_{i=1}^{n} x_i \cdot M_i}} \tag{12}$$

where σ_a is the alloy surface tension in $(N.m^{-1})$, η_a is the alloy viscosity in $(Pa.s)$, x_i is the molar fraction of the i alloy component, n is the total number of components of the alloy, and M_i is the molar weight of the alloy component i.

Substituting Eq. (12) into Eq. (6), we get,

$$\sigma_a\left(T, x_i, x_j, \cdots, x_n\right) = \frac{15}{16} \cdot \frac{h \cdot N_{Av}}{\sum_i x_i \cdot V_i + \Delta V^E} \cdot \sqrt{\frac{\overline{R} \cdot T}{\sum_{i=1}^{n} x_i \cdot M_i}} \cdot \exp\left[\frac{\sum_i x_i \cdot \Delta G_i^* - \alpha \cdot \Delta H}{\overline{R} \cdot T}\right] \tag{13}$$

4. Results and discussion

Figure 1 presents the viscosity of pure aluminum as a function of temperature, simulated by Kaptay [7] theoretical model compared to experimental data [37]. The model fits the experimental scatter well, but a correction in the melting temperature for pure aluminum was carried out according to the model description provided in [7], although, for the case of Al, no correction was applied by the author [7]. **Figure 2** shows the evolution of viscosity of molten Cu as a function of temperature, where the theoretical model is compared to experimental results [37], where a deviation is noticed for temperatures close to the melting temperature.

In **Figure 3**, the viscosity of pure molten silicon is plotted against temperature. As mentioned, the melting temperature of Si has been assumed as $T_{m\ corr,\ Si} \cong 596.85$ K [22], which is less than the correction of melting temperature of pure Si proposed by Kaptay [7], that is, $T_{m\ corr,\ Si} \cong 870$ K. This correction was carried out by the author for some pure molten metals, so that the model could also fit the experimental scatter of these elements. As can be observed, a correction in the melting temperature was needed not only for Si, Ge, Sb, and Bi but also for pure molten Al, based on the same set of experimental data.

In **Figure 4**, it can be seen that Kaptay's model [7] well the experimental data well for the viscosity of molten magnesium as a function of temperature [38–40].

In **Figures 5–8**, the surface tension for pure Al, Cu, Si, and Mg is depicted as a function of temperature, calculated from surface tension-viscosity relations according to Egry [34] and Kaptay [7] formulations. For all cases, except for Si in **Figure 7**, the surface tension exhibits a trend to decrease as the temperature increases. It also can be noticed that for all cases, they diverge close to the melting point. For high temperatures, both models yield very close results. The best agreement observed between the two models is that of pure Mg, as shown in **Figure 8**.

Figures 9–12 show the evolution of viscosity as a function of temperature for all examined alloys, where it can be seen that the two models generally exhibit similar

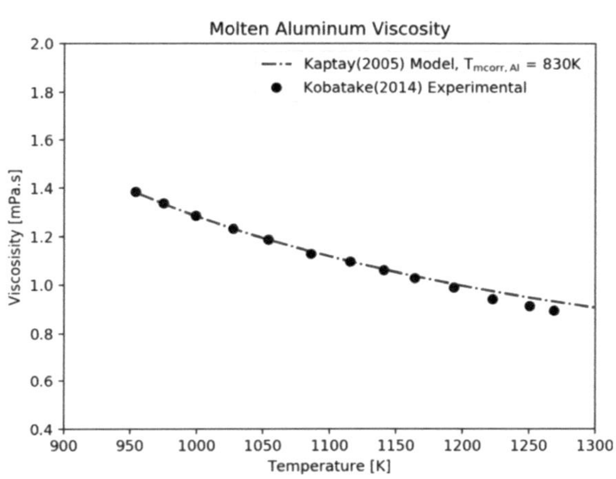

Figure 1.
Calculated viscosities of the pure molten Al as a function of temperature compared to experimental data [37].

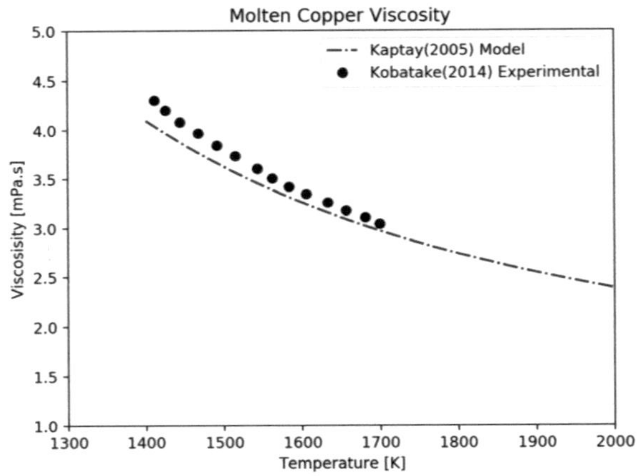

Figure 2.
Calculated viscosities of the pure molten Cu as a function of temperature compared to experimental data [37].

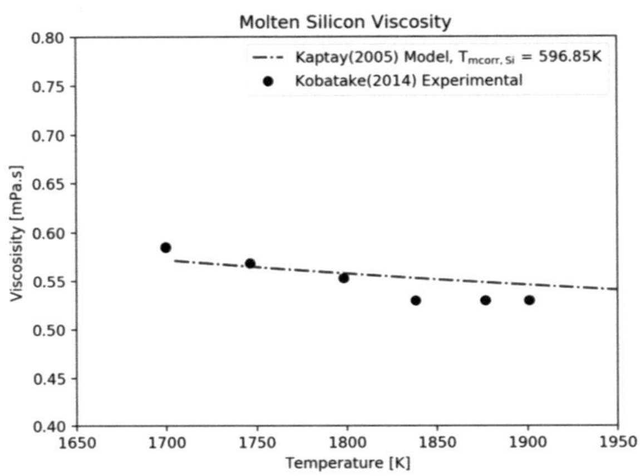

Figure 3.
Calculated viscosities of the pure molten Si as a function of temperature compared to the experimental data [37].

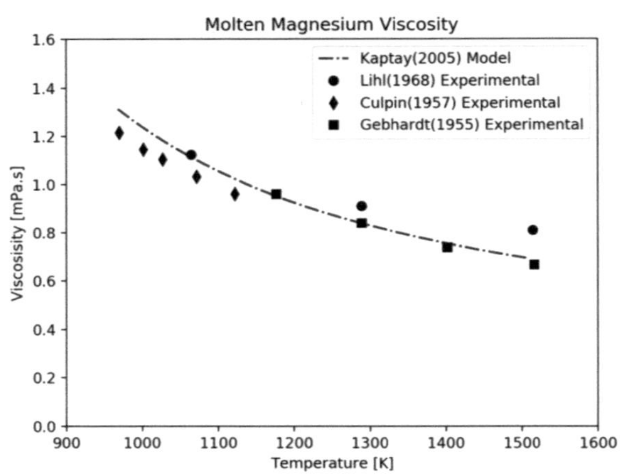

Figure 4.
Calculated viscosities of the pure molten Mg as a function of temperature compared to the experimental data.

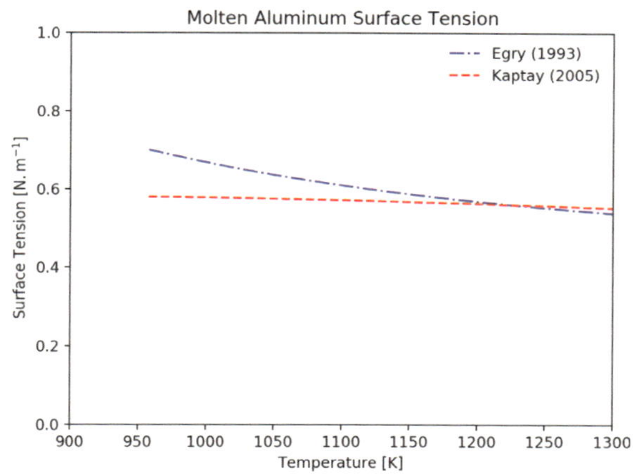

Figure 5.
Comparison between surface tension of pure molten Al as a function of temperature provided by both Egry's and Kaptay's surface tension-viscosity relation models.

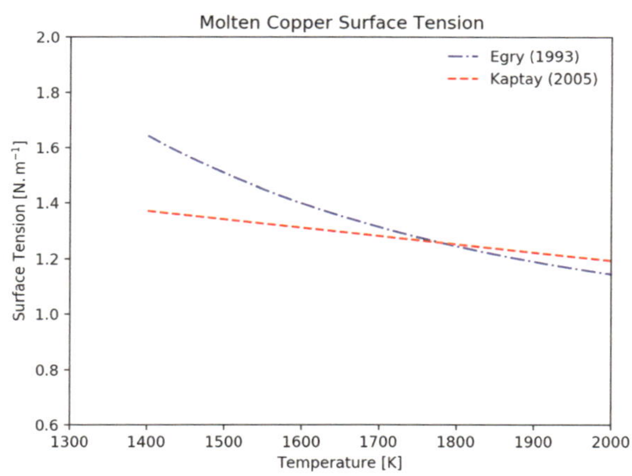

Figure 6.
Comparison between surface tension of pure molten Cu as a function of temperature provided by both Egry's and Kaptay's surface tension-viscosity relation models.

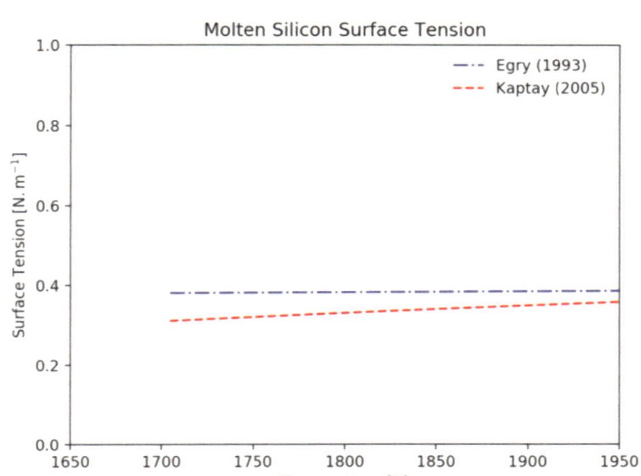

Figure 7.
Comparison between surface tension of pure molten Si as a function of temperature provided by both Egry's and Kaptay's surface tension-viscosity relation models.

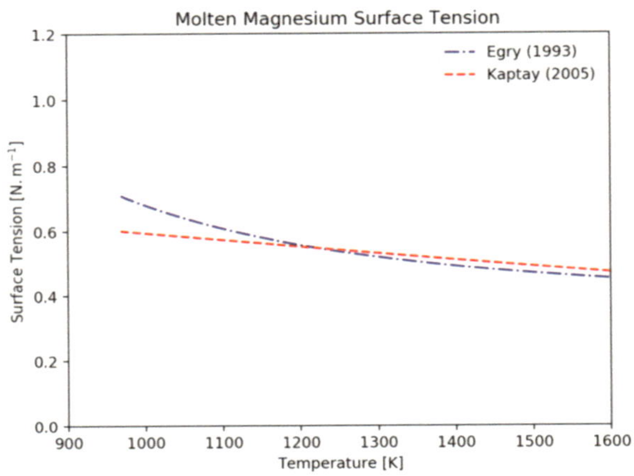

Figure 8.
Comparison between surface tension of pure molten Mg as a function of temperature provided by both Egry's and Kaptay's surface tension-viscosity relation models.

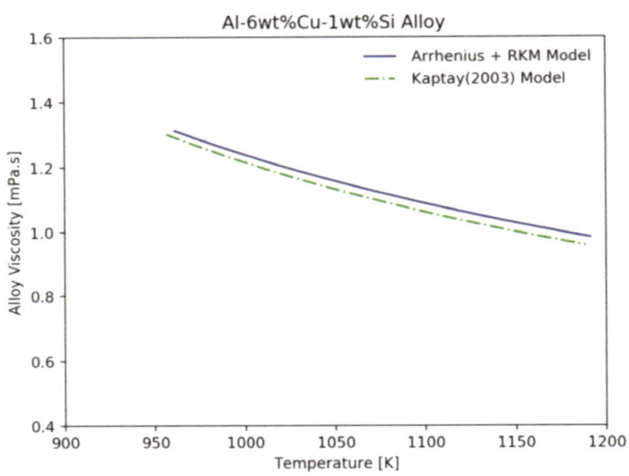

Figure 9.
Simulations of viscosity of Al-6wt%Cu-1wt%Si as function of temperature: Arrhenius-type equation and Kaptay model.

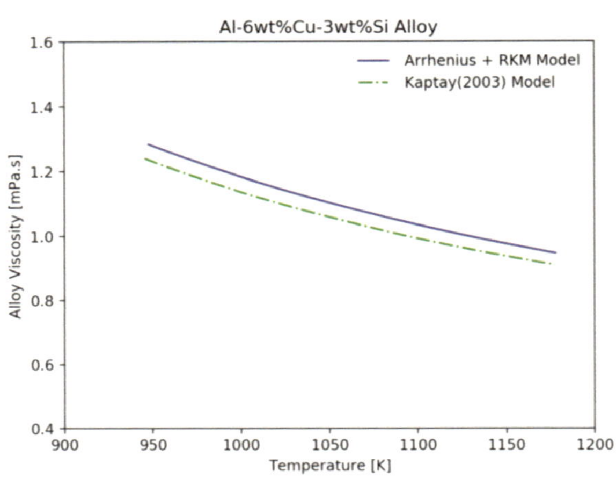

Figure 10.
Simulations of viscosity of Al-6wt%Cu-3wt%Si as function of temperature: Arrhenius-type equation and Kaptay model.

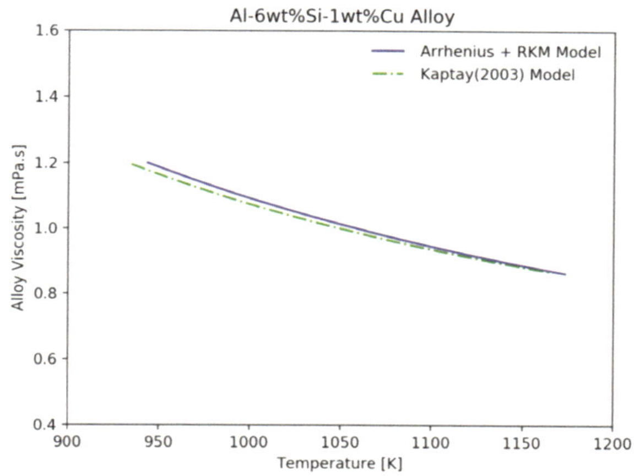

Figure 11.
Simulations of viscosity of Al-6wt%Si-3wt%Cu as function of temperature: Arrhenius-type equation and Kaptay model.

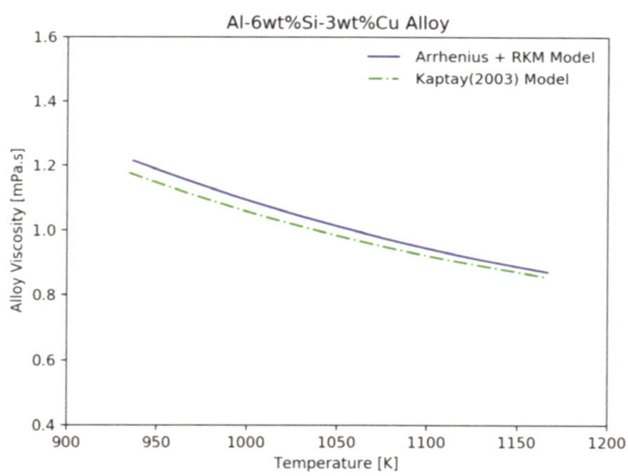

Figure 12.
Simulations of viscosity of Al-6wt%Si-3wt%Cu as function of temperature: Arrhenius-type equation and Kaptay model.

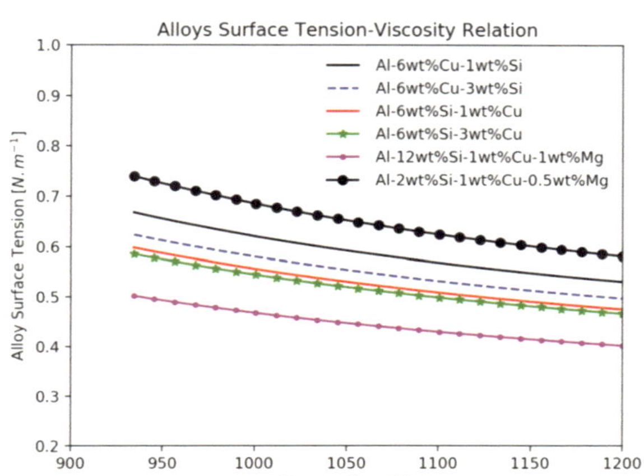

Figure 13.
Application of derived surface tension-viscosity relation for Al-Cu-Si ternary alloys and quaternary Al-Cu-Si-Mg alloys as function of temperature.

results. The greatest deviation between the models can be observed for the Al-6wt% Cu-3wt%Si alloy, **Figure 10**; but even in this case, a relatively good agreement can be considered. This may be related to the non-ideal term of viscosity of the Redlich-Kister equation, whose coefficients do not depend on the alloy composition but depend on the temperature [36].

Figure 13 shows the application of the derived surface tension relation equation for alloys, Eq. (13), to ternary Al-Cu-Si and quaternary Al-Cu-Si-Mg alloys. As can be noticed, the increase in the alloy Si content decreases the surface tension. The lowest surface tension profile is associated with the alloy having the highest Si content, that is, for the Al-12wt%Si-1wt%Cu-1wt%Mg quaternary alloy.

5. Conclusion

A solution for viscosity of alloys has been derived based on an extension of Egry's surface tension-viscosity relation equation for pure metals [34] and Kaptay's [10] unified solution for viscosity of multicomponent alloys. Kaptay and Egry's surface tension-viscosity relations were plotted against temperature. It was shown that for the pure metals analyzed in the present study, for lower temperatures, both relations deviated from each other. However, for higher temperatures, a better agreement has been achieved. The derived solution for surface tension was plotted against temperature for ternary Al-Cu-Si and Al-Cu-Si-Mg alloys. It was shown that with increasing Si alloy content, the surface tension of Al-based alloys decreases, which is in agreement with the casting practice of Aluminum-based alloys.

Acknowledgements

The authors acknowledge the financial support provided by FAPERJ (The Scientific Research Foundation of the State of Rio de Janeiro), CAPES, and CNPq (National Council for Scientific and Technological Development).

Author details

Ivaldo Leão Ferreira[1]*, José Adilson de Castro[2] and Amauri Garcia[3]

1 Faculty of Mechanical Engineering, Federal University of Pará, UFPA, Belém, PA, Brazil

2 Graduate Program in Metallurgical Engineering, Fluminense Federal University, Volta Redonda, RJ, Brazil

3 Department of Manufacturing and Materials Engineering, University of Campinas —UNICAMP, Campinas, SP, Brazil

*Address all correspondence to: ileao@ufpa.br

References

[1] Dinsdale AT, Quested PN. The viscosity of aluminium and its alloys—A review of data and models. Journal of Materials Science. 2004;**39**:7221-7228

[2] Brooks RF, Dinsdale AT, Quested PN. The measurement of viscosity of alloys—A review of methods, data and models. Measurement Science and Technology. 2005;**16**:354-362

[3] Iida T, Shiraishi Y. In: Kawai Y, Shirashi Y, editors. Chapter 4: Viscosity Handbook of Physico-Chemical Properties at High Temperatures. Japan: Iron and Steel Institute; 1988

[4] Hildebrand JH. Viscosity and Diffusivity: A Predictive Treatment. New York: John Wiley & Sons Inc; 1977

[5] Walther C. The evaluation of viscosity data. Erdöl und Teer. 1931;**7**: 382-384

[6] Chhabra RP, Tripathi A. A correlation for the viscosity of liquid metals. High Temperatures - High Pressures. 1993;**25**:713-718

[7] Kaptay G. A unified equation for the viscosity of pure liquid metals. Zeitschrift fuer Metallkunde. 2005;**96**: 1-8

[8] Zivkovic D, Manasijevic D. An optimal method to calculate the viscosity of simple liquid ternary alloys from the measured binary data. Calphad. 2005;**29**:312-316

[9] Kucharski M. The viscosity of multicomponent systems. Zeitschrift fuer Metallkunde. 1986;**77**:393-396

[10] Kaptay G. A new equation to estimate the concentration dependence of the viscosity of liquid metallic alloys from the heat of mixing data. In: Proceedings of microCAD 2003 Conference. Section Metallurgy, University of Miskolc; 2003. pp. 23-28

[11] Zhong XM, Liu YH, Chou KC, Lu XG, Zivkovic D, Zivkovic Z. Estimating ternary viscosity using the thermodynamic geometric model. Journal of Phase Equilibria. 2003;**24**: 7-11

[12] Singh RN, Nadan R, Sommer F. Viscosity of liquid alloys: Generalization of Andrade's equation. Monatshefte für Chemie. 2012;**143**:1235-1242

[13] Jeyakumar M, Hamed M, Shankar S. Rheology of liquid metals and alloys. Journal of Non-Newtonian Fluid Mechanics. 2011;**166**:831-838

[14] Solek K, Korolczuk-Hejnak M, Slezak W. Viscosity measurement for modeling of continuous steel casting. Archives of Metallurgy and Materials. 2012;**57**:333-338

[15] Moelwyn-Hughes EA. Physikalische Chemie. Stuttgart: Thieme; 1970

[16] Iida T, Ueda M, Morita Z. On the excess viscosity of liquid alloys and the atomic interaction of their constituents. Tetsu to Hagane. 1976;**62**: 1169-1178

[17] Kozlov LY, Romanov LM, Pod'yachev AP, Petrov NN, Kostygova OV. Calculation of the viscosity of multicomponent iron- and nickel-base melts. Izvestiya Vysshikh Uchebnykh Zavedenii, Chernaya Metallurgiya. 1982; **5**:1-4

[18] Seetharaman S, Du SC. Estimation of the viscosities of binary metallic melts using Gibbs energies of mixing. Metallurgical and Materials Transactions B: Process Metallurgy and Materials Processing Science. 1994;**25B**: 589-595

[19] Hirai M. Estimation of viscosities of liquid alloys. ISIJ. 1993;**33**(2):281-285

[20] Budai I, Benkö MZ, Kaptay G. Comparison of different theoretical models to experimental data on viscosity of binary liquid alloys. Materials Science Forum. 2007;**537–538**: 489-496

[21] Cheng J, Gröbner J, Hort N, Schmid-Fetzer R. Measurement and calculation of the viscosity of metals—A review of the current status and developing trends. Measurement Science and Technology. 2014;**25**:062001. 9pp

[22] Butler JAV. The thermodynamics of the surfaces of solutions. Proceedings of the Royal Society of London A. 1932;**135**: 348-375

[23] Tanaka T, Lida T. Application of a thermodynamic database to the calculation of surface. Tension for iron-base liquid alloys. Steel Research International. 1994;**65**:21-28

[24] Hoar TP, Melford DA. The surface tension of binary liquid mixtures: Lead + tin and lead + indium alloys. Transactions of the Faraday Society. 1957;**53**:315-326

[25] Tanaka T, Hack K, Lida T, Hara S. Application of thermodynamic databases to the evaluation of surface tension of molten alloys, salt mixture and oxide mixtures. Zeitschrift fuer Metallkunde. 1996;**87**:380-389

[26] Fowler RH. A tentative statistical theory of Macleod's equation for surface tension, and the parachor. Proceedings of the Royal Society of London A. 1937; **159**:229-246

[27] Shih W-H, Stroud D. Theory of the surface tension of liquid metal alloys. Physical Review B. 1985;**32**:804-811

[28] Aqra F, Ayyad A, Takrori F. Model calculation of the surface tension of liquid Ga-Bi alloy. Applied Surface Science. 2011;**257**:3577-3580

[29] Goicoechea J, Garcia-Cordovilla C, Louis E, Pamies A. Surface tension of binary and ternary aluminium alloys of the systems Al-Si-Mg and AI-Zn-Mg. Journal of Materials Science. 1992;**27**: 5247-5252

[30] Egry I. Surface tension of compound forming liquid binary alloys: A simple model. Journal of Materials Science. 2004;**39**:6365-6366

[31] Amore S, Brillo J, Egry I, Novakovic R. Surface tension of liquid Cu–Ti binary alloys measured by electromagnetic levitation and thermodynamic modelling. Applied Surface Science. 2011;**257**:7739-7745

[32] Egry I. On a universal relation between surface tension and viscosity. Scripta Metallugica et Materialia. 1992; **26**:1349-1352

[33] Born M, Green HS. A general kinetic theory of liquids III. Dynamical properties. Proceedings of the Royal Society of London A. 1947;**190**:455-474

[34] Egry I. On the relation between surface tension and viscosity for liquid metals. Scripta Metallugica et Materialia. 1993;**28**:1273-1276

[35] Andrade EN da C. The theory of the viscosity of liquids: I. Philosophical Magazine. 1934;**17**:497-511

[36] Zhang F, Du Y, Liu S, Jie W. Modeling of the viscosity in the Al-Cu-Mg-Si system: Database construction. Calphad. 2015;**49**:79-86

[37] Kobatake H, Schmitz J, Brillo J. Density and viscosity of ternary Al-Cu-Si alloys. Journal of Materials Science. 2014;**49**:3541-3549

[38] Lihl F, Nachtigall E, Schwaiger A. Viskositätsmessungen an binären Legierungen des Aluminiums mit Silizium, Zink, Kupfer und Magnesium.

Zeitschrift fuer Metallkunde. 1968;**59**:
213-219

[39] Culpin MF. The viscosity of liquid magnesium and liquid calcium. Proceedings of the Physical Society. 1957;**70**:1079-1086

[40] Gebhardt E, Becker M, Dorner S. Effect of alloying additions on the viscosity of aluminum. Aluminum. 1955; **31**:315-321

Chapter 8

Natural and Artificial Superwettable Surfaces-Superficial Phenomena: An Extreme Wettability Scenario

Cristina Elena Dinu-Pîrvu, Roxana-Elena Avrămescu, Mihaela Violeta Ghica and Lăcrămioara Popa

Abstract

With the help of biomimetics, superficial characteristics were transposed, through various methods, onto artificially obtained materials. Many industrial fields applied surface architecture modifications as improvements of classic materials/methods. The medico-pharmaceutical, biochemical, transportation, and textile fields are few examples of industrial areas welcoming a "structural change." Anti-bioadhesion was widely exploited by means of antibacterial or self-cleaning fabrics and cell culturing/screening/isolation. Anti-icing, antireflective, and anticorrosion materials/coatings gained attention in the transportation and optical device fields. Interdisciplinary approaches on extreme wettability include "solid-fluid" formations called liquid marbles, which will be further discussed as a superhydrophobic behavior exponent.

Keywords: superficial phenomena, extreme wettability, special surface architecture, liquid marble

1. Introduction

Since ancient times, humans observed special features which helped plants and animals survive in harsh environments. These properties were unraveled with the help of microscopical investigative techniques, which led to a more thorough understanding of superficial phenomena. Natural unique superficial architectures, like the lotus and rose petal effect, became iconic. Empirical models of wettability were developed (Young, Cassie, and Wenzel) to fully explain the behavior of liquids in contact with special surfaces.

Along with the fulminant expansion of technology during the last decades, a huge progress was also registered in surface sciences. Microscopical analysis techniques revealed surfaces' special architectures and solved many mysteries regarding plants and animals' adaptation to harsh environments (e.g., Namib beetles' survival in the desert). Extreme wettability (superhydrophobicity/superhydrophilicity) was assigned to many natural phenomena, such as raindrops not collapsing while dropping onto ash-covered soil and moss storing the exact amount of water needed to survive. In particular, superhydrophobic surfaces which display a contact angle higher than 150°, a sliding angle smaller than 10°, and no hysteresis attracted researchers' attention. Apart from theoretical aspects on wettability, which are a part of the paper, natural extreme wettability models will be discussed (lotus leaf, rose petal, and insect wings).

Principles of biomimetics are included in this chapter, as special superficial properties were adapted to human necessities and used as a model in many industrial areas, including nanotechnologies. Biomimetics is, in this case, "the thread that makes the dress complete," or, in other words, "the scene that completes the movie." Applications related to superhydrophobicity will be presented: development of self-cleaning and low friction surfaces, satellite antennas, solar and photovoltaic panels, exterior glass, etc. Studies concerning superhydrophobic surfaces' applications in various domains will also be submitted: prevention of bacterial adhesion, of metal corrosion, of surface icing in humid atmosphere and low-temperature conditions, blood type determination techniques, etc. Efficient, cost-effective, ecological, and reproductible methods are still developing so that mass production of quality materials becomes a fact.

The chapter will also bring into attention an important exponent of superhydrophobicity: special structures called liquid marbles. The unique "solid-fluid" formations are regarded as soft objects, due to the microliter droplet encapsulated in hydrophobic particles. Practical uses include micro-reactors, miniature cell culturing, or screening devices, successfully replacing classical methods with cost- and reactive-efficient, low-toxicity analysis techniques. Other properties will be submitted along with applications in various fields.

2. Biomimetics: biology vs. technology

As human kind evolved, passing through the test of time, many necessities turned out as a result of convenience in everyday life activities. Thus, classical materials like wood, metal, and ceramic became no longer suitable and efficient, lacking performance in many domains (e.g., pharmaceutical, medical devices, weaponry, etc.). Aiming a more complex approach on artificial materials, the concept "materials by design" came to life (Bernadette Bensaude-Vincent, 1997) [1]. The concept refers to developing "composite" materials, which reunite properties of already known ones: heat resistance and time durability of ceramics, lightness of plastic, hardness, and breaking resistance of metals. Depending on quality requirements of the final product, many design possibilities came out, exhibiting improved sustainability, cost-effectiveness, durability, and an environmentally friendly character.

Undoubtedly, a much older concept, "biomimetics," also led to obtaining performant structural materials and is intricated to the "materials by design" concept. The term itself ("biomimetics") was firstly introduced by Otto Schmitt [1], but its principles are considered to be used since Leonardo da Vinci (1452–1519) while designing flying machines after analyzing bird's ability to fly [2].

According to some beliefs, biomimetics is a transfer of ideas between biology and technology, aiming to obtain superior device. A more complex approach refers to it as being "a study of the formation, structure or function of biologically produced substances, materials, biological mechanisms and processes especially for the purpose of synthetizing similar products by artificial mechanisms which mimic natural ones" [2].

As expected, a lot of controversial interpretations arise from different approaches of biology and engineering, considered "baselines" of biomimetics. On the one hand, biology relates to living organisms (cells, plants, and animals) which evolve following a natural DNA-embedded cycle. On the other hand, engineering relies on human intelligence which develops successive steps in order to obtain a final product [1].

Conflicts between biology and technology interpretations are classified through a Russian problem-solving system TRIZ (*Teorija Reshenija Izobretatel'skih Zadach*—"Theory of Inventive Problem Solving"), 40-standard features which offer solutions from both perspectives. For example, a few characteristics such as "keep poison out," "self-cleaning," "surface properties," and "waterproof" became conflict nr. 30-"external harm affects the object" [1].

Therefore, it became appropriate to say that "superficial properties" is a concept which can be regarded in the light of both "biology" and "technology." Structural investigation of natural (biological) surfaces is performed using microscopical techniques. Technical interpretations of these surfaces and empirical models arise. Examples include special wettable surfaces like the lotus leaf, rose petal, Salvinia leaf, insect eyes, wings, fish scales, etc. They provide templates in designing new engineered materials exhibiting improved properties compared to classical materials. Such artificially obtained materials and coatings can be considered results of "materials by design" and "biomimetics" concepts, as a reunion of biological inspiration and human engineering. Even though many contradictory assessments take place, it is important to state that biology and technology functioned perfectly together when inventing the Velcro closure system according to the way burdock (*Arctium* sp.) spreads its seeds, the helicopter inspired by the body of the dragonfly, the submarine resembling a whale, etc.

3. Extreme wettability: special patterns

3.1 Understanding wettability

As is well known, surface wettability characterizes interfacial phenomena between a liquid and a solid support. The liquid's behavior on the studied surface is in fact an indicator of wettability, a superficial property which helps evaluate hydrophilicity/hydrophobicity of a solid. The quantitative indicator of wettability is represented by the contact angle, given by Young's equation (Eq. (1)):

$$\cos\theta = \frac{\gamma_{SV} - \gamma_{SL}}{\gamma_{LV}} \tag{1}$$

where θ is the contact angle, γ_{SV} is the solid-vapor superficial energy, γ_{SL} is the solid-liquid superficial energy, and γ_{LV} is the liquid-vapor superficial energy [3].

The equation establishes an equilibrium between superficial energies at the solid-liquid-air interface. However, adaptations of Young's equation were proposed by Wenzel [4] and Cassie-Baxter [5], after it was proven that the original equation only applies to homogenous, smooth surfaces and that the contact angle is influenced by the support's rugosities, as a surface roughness indicator.

Wenzel's equation (Eq. (2)) applies to non-smooth surfaces. Surface rugosity is interpreted through the roughness factor r, defined as ratio of the actual rough surface area to the geometric area projected on a relatively smooth surface. This adapted equation refers to an apparent contact angle θ', as follows (Eq. (2)):

$$\cos\theta' = \frac{r(\gamma_{SV} - \gamma_{SL})}{\gamma_{LV}} = r\cos\theta \tag{2}$$

Another relationship defining an apparent contact angle θ' is similar to Wenzel's equation, with the difference that the surface's rugosities are separated by impenetrable air pockets (Cassie-Baxter wetting model). The surface f in direct contact with the liquid is considered, as follows (Eq. (3)):

$$f = \frac{\Sigma a}{\Sigma (a + b)} \tag{3}$$

where a and b are the contact areas with the drop (a) and, respectively, air (b). Considering (1 − f) the drop-air contact area and a contact angle of 180°, the calculation formula corresponding to the Cassie-Baxter wetting regime is shown in Eq. (4):

$$\cos\theta' = f\cos\theta + (1 - f)\cos 180^\circ = f\cos\theta + f - 1 \tag{4}$$

Empirical models of the Young, Wenzel, and Cassie-Baxter wetting states are presented in **Figure 1**.

Other interpretations by Quéré et al. [6, 7] consider the Wenzel wetting regime as an equilibrium state of the Cassie model: a critical value of the fraction f determines a critical contact angle θ_c, determined by the following equation (Eq. (5)):

$$\cos\theta_c = \frac{1 - f}{r - f} \tag{5}$$

Since wettability studies continue to unfold, researchers recently proved that Wenzel and Cassie wetting regimes actually co-occur on the same support surface. Hydrophobic surfaces with linear or pillar patterns exhibit both a Cassie levitating state corresponding to drops placed on the support and also a Wenzel pinned state for drops which come into contact with the surface after the impact. Transitions between these states were also reported as a result of external stimuli influence [8, 9]. Wenzel to Cassie and Cassie to Wenzel transitions were analyzed through sequential squeezing and releasing between texture surfaces of nonadhesive plate. Results indicate that both regimes exist at the same time on a double-scaled textured surface, resembling natural micro- and nano-surface architecture: the Wenzel state is characteristic for the larger texture and Cassie to the smaller one [9]. Further investigations consisted in exploiting these characteristics and developing super-repellent materials, also based on natural models, following the principles of biomimetics.

The "superwettability system," briefly presented in **Figure 2**, includes a much extensive approach on wetting states, depending on the liquid type, the solid support's architecture, and the environment in which the phenomenon is described. Thus, the terms discussed above (hydrophilicity/hydrophobicity) refer to water's behavior in air and upon flat surfaces. Regarding low-surface liquids, such as oils, the "oleophilic/oleophobic" concepts are defining. Moreover, if the support

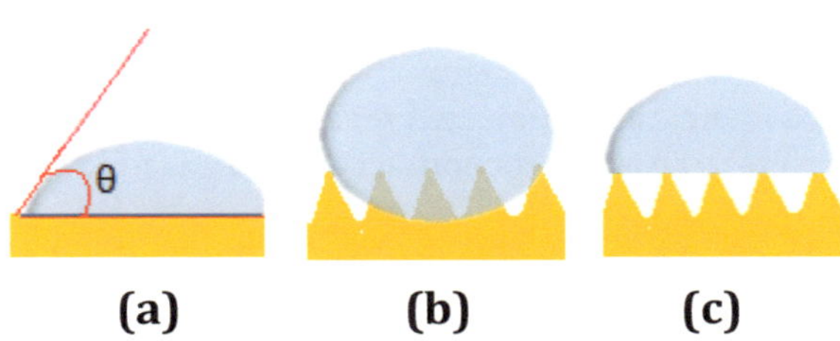

Figure 1.
Comparison between wetting regimes: (a) Young, (b) Wenzel, and (c) Cassie.

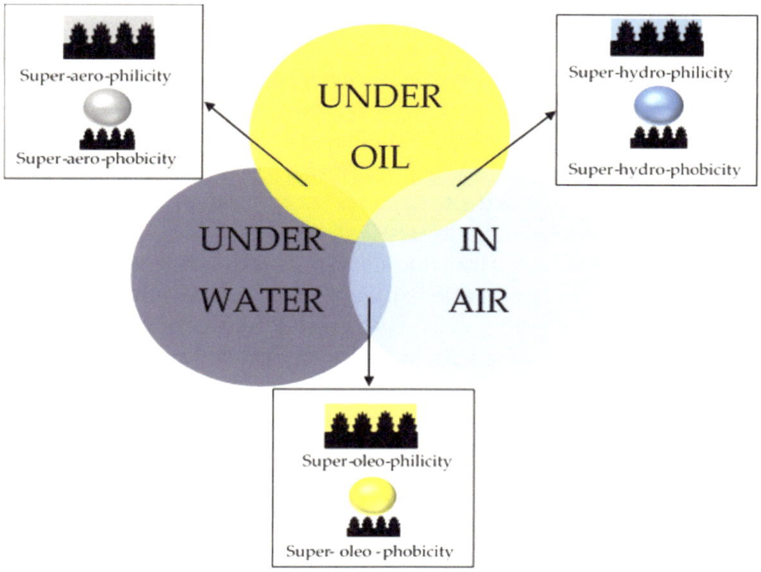

Figure 2.
The "superwettability" system.

exhibits a nano-/micro-rough architecture, then the behavior of liquids when contacting such a surface is known as "superhydrophobic/superhydrophilic" and "superoleophobic/superoleophilic." Corresponding wetting behaviors under water for structured rough supports are known as "superoleophobic/superoleophilic" and "superaerophobic/superaerophilic." If placed under oil, then the appropriate approach refers to "superhydrophobicity/superhydrophilicity" and also "superaero-phobicity/superaerophilicity" [10].

3.2 Natural designs

Natural special surfaces transposed as survival skills in animals and plants captivated attention of researchers. They investigated and applied in practice what nature provided. Apart from scientists, novelists like Jules Verne were fascinated by certain elements from the environment and used them as inspirational sources to imagine innovative devices, mostly designed as transformational means and considered eccentrical in that era: the Nautilus submarine whose shape resembled a whale, the eponymous Steam House—a mechanical elephant, the helicopter imagined starting from insects' flight mechanisms and shapes, etc.

Apart from mechanical devices, natures' kingdom offered humans the possibility to improve artificial materials, based on the evolution of SEM analysis techniques in the 1960s. Detailed investigations of surface structure and properties were performed. As a result, surface architecture was held responsible for many phenomena which were not explained at that time: how plants maintain clean in marshy environments and how their water needs are satisfied during high-temperature exposure. From this category, two types of surface structures, designed as micro- and nano-scaled patterns, confer superhydrophobicity to the leaves of certain plant species: lotus, rice, and taro. Another model which confers water repellency was attributed to a unitary structure of 1–2 μm fibers (Chinese watermelon, Ramee leaves). Also, vertical/horizontal hairs were attributed in the property of water repellency in case of *Alchemilla vulgaris* and, respectively, *Populus* sp. [11–13].

The iconic plant superhydrophobic behavior belongs to the lotus leaf (*Nelumbo nucifera*). Surface wettability of the leaves is considered to be derived from the Cassie-Baxter wetting model: convex micrometric papillae along with nanometric wax needles determine water contact angles higher than 150°. This special architecture, also known as "The Lotus Effect," allows dust particles and other impurities to be collected by raindrops while rolling off (**Figure 3**) [14].

The other side of the lotus leaf presents no waxy crystals, but has tabular nano-groove convex lumps which confer inverse wettability [15].

An example of unique structural characteristics is the carnivorous plant *Nepenthes alata*. Its prey is caught due to oleophobic features which allow insects to slide down to the digestive cavity, capturing them [16]. Contributing to the survival of the *Cladonia chlorophaea* lichen are hydrophobic strains ending in cup-shaped structures which limit water storage, preventing excessive accumulation and further damage to the plant [17].

An exponent of plant adaptation to harsh environment conditions is represented by *Salvinia molesta*, the water fern, who gave rise to "The Salvinia Paradox": hydrophobic hairs ending in hydrophilic peaks retain an air film while submerged, allowing respiration [18]. The thin air film retained at the air-water interface also enables *Oryza sativa* (rice) to carry on photosynthesis, enhancing gas exchange and diminishing Na^+ and Cl^- intrusion through submerged leaves through salt-water floods [19, 20].

"The Rose Petal Effect" reveals how nano-folds covered with micro-papillae of rose petals confer contact angle values of 152°, resembling the Cassie impregnating model: water droplets maintain their spherical shape, adhere to the surface, and do not slip when turned upside down. Compared to the lotus leaf architecture, this wetting regime is characterized by a liquid film which impregnates the papillae, leaving only some dry areas. A dependence was observed between the drops' volume and surface tension: the equilibrium is ruined and the drop falls if it exceeds 10 μL in volume. Thus, smaller drops stay stable, while raindrops slide off, since they are bigger [21]. **Figure 4** illustrates a comparison between the rose petal (a) and the lotus leaf (b) surface structure [22].

Transitioning from the plant to the animal kingdom, it is important to state that apart from "slippery" surfaces discussed above, "adherent" superhydrophobic surfaces were also noted: the gecko lizard's finger structures confer them the ability to climb even perfectly vertical walls, due to micrometric lamellae divided into nanometric setae. A drop placed on this surface retains its shape even in an antigravity

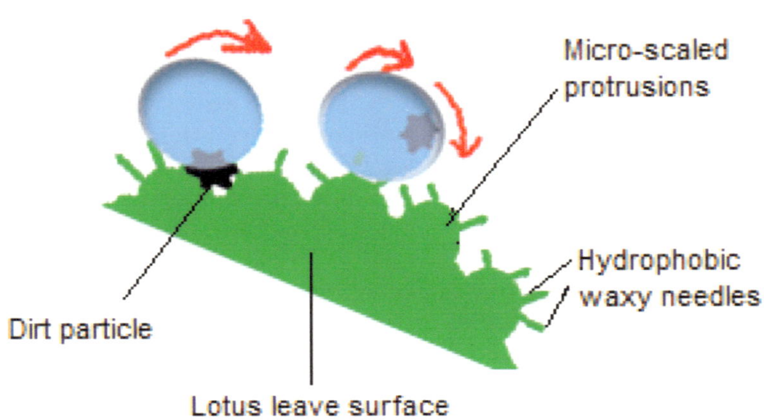

Figure 3.
Lotus leaf structure. Dirt particle removed by rain.

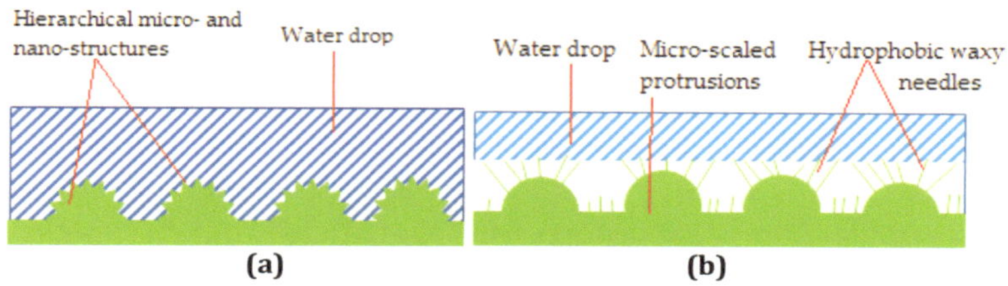

Figure 4.
(a) Rose petal surface structure (Cassie impregnating wetting state) and (b) lotus leaf surface structure (Cassie state) [22].

position [23, 24]. The gecko feet model inspired climbing a glass building using Kevlar and polyurethane special gloves [25]. The group of adhesive superhydrophobic natural surfaces includes also the rose petals, as previously discussed.

Regarded at first from a different angle, the insects' ability to fly, to maintain impurity and water-free wings, was later attributed to superhydrophobicity. Microscales hierarchically disposed on insect wings are responsible for maintaining them dry (**Figure 5**) and also exhibit, in some cases, antibacterial activity (cicada wings are bactericidal against Gram-negative bacteria) [26, 27].

For some insects, patterns joining superhydrophobicity in alternation with superhydrophilicity represent an adaptation to harsh environmental conditions: *Stenocara gracilipes* (the Namib desert beetle) shows on its back a real storage system which captures water from atmospheric humidity, due to superhydrophobic waxy edges and superhydrophilic peaks [17]. In relation with survival skills, *Argyroneta aquatica* (the diving spider) creates around itself a hydrophobic artificial lung, which is permeable to gases and allows underwater living [10].

3.3 Engineered superwettability—materials and coatings: practical applications

Moving on from the theoretical field, extreme wettability is regarded an open gate for numerous everyday life and also industrial applications.

Following biomimetic principles and varying surface templates, innovative materials are fabricated, depending on qualitative requirements. The first artificial superhydrophobic materials appeared in the early 1990s: the submicrometer-roughed glass plates hydrophobized with fluoroalkyl trichlorosilane (CA = 155°) [28], fractal surfaces covered in n-alkyl ketene (CA = 174°) [29, 30], and ion-plated polytetrafluorethylene (PTFE) coatings with nanometric rugosities [31]. In the 2000s, surface topography studies were correlated with surface chemistry, leading to patterned silicone surfaces with low wettability [32].

Figure 5.
Insect wing-microscaled superhydrophobicity.

Techniques used to confer surface roughness are still improving, along with transparency, permeability, resistance, and color change imparting methods [33–35]. Nature proved that hierarchical surface structures are responsible for surface special wettability, and not fluorocarbon derivatives, as it was considered at that time [36].

The most popular known procedures used to artificially obtain superhydrophobic surfaces include chemical reactions in a humid atmosphere [37], thermic reactions [38], electrochemical deposition [39], individual/layer-by-layer assembling [40], etching [41], chemical vapor deposition [42], and polymerization reactions [43]. Substrates include glass, metals (Cu, Ti, Zn), and cotton, and resulted structures exhibit CA > 150°, mimicking natural patterns [44]. For example, the rose petal was used as template in order to obtain polymeric coatings, resulting in "adhesive" superhydrophobicity [21]. Patterns resembling surface design of the lotus leaf were also fabricated through eco-friendly methods, without toxic solvents [45].

Fluorocarbon and silicone derivatives were preferred as substrates in fabricating superhydrophobic surfaces, assuming that the larger the number of flor atoms, the higher is the hydrophobicity [46]. Nowadays, these materials are replaced with biodegradable ones, such as agricultural residues [47]. Recent studies indicate that lignocellulose can be successfully used in obtaining fire-proof coatings [48]. Another example of eco-friendly superhydrophobic coatings includes waterborne resins from aqueous silanes and siloxane solutions with silica nanoparticles applied as protective coatings to cultural heritage (marble, sandstone, cotton, ceramic artifacts) [49].

The *Slippery Liquid-Infused Porous Surface* (SLIPS) technology includes smooth coatings applied onto military uniforms and medical gowns in order to avoid biological fluid contamination, due to surface fluids incorporated into a micro-/nano-porous substrate [50, 51]. The field of *anti-bioadhesion* also involves protein adsorption, bacterial adhesion, and cell culturing media, all of them wettability-dependent phenomena [52]. Thus, in vitro studies regarding platelet adhesion on implants reveal that no adhesion happens on TiO_2 nanotube-covered supports. Moreover, polydimethylsiloxane (PDMS) surfaces with various sized-rugosities, superposed scale plates, submicron structures, and nanostructured and smooth surfaces, proved the highest effectiveness against blood platelet adhesion in superposed scale plate surface [53]. Antibacterial cellulose fibers modified with siloxanes and silver nanoparticles show durable activity against *Escherichia coli* and *Staphylococcus aureus* [54], while bactericidal action against *Pseudomonas aeruginosa* was discovered for fluoroalkyl silane-hydrophobized glass [55].

The result of joining extreme wettability surfaces are patterns which promote development of cells planted as hydrogels/solutions in the hydrophilic zone. Advantages of the method include lack of lateral contamination risks due to hydrophobic separative borders, efficiency, economic analysis method, the possibility of real-time screening, and noninvasive diagnosis [56–58].

Other high-impact applications of superhydrophobic surfaces include *anti-icing*, *antireflective*, and *low friction* properties, mostly popular in the marine and aviation transportation fields, mirrors, and lens industry [59–62]. These properties along with a low adhesion degree, a high contact angle, and a low sliding angle allow impurity collection, while rolling off represent desired characteristics for windshields, exterior windows, and solar panels [63].

Since *metal corrosion* is a contemporary problem, superhydrophobic anticorrosion treatments were developed: coating techniques (microwave chemical vapor deposition, followed by immersion) with fluorochloride silanes of magnesium alloys and substrate modifications (Al with hydroxides, Zn immersed in superhydrophobic solutions), proving resistance against acids, alkaline, or saline solutions [64].

Closely following the corrosion issue is *friction reduction*, which is of interest in aeronautics and ships. The shark skin and the lotus leaf are models in designing continuous surface films with self-contained air bubbles, able to reduce laminar and turbulent liquid flow, lowering friction forces. Moreover, recent progress includes high-pressure-resistant special surfaces, with a high impact in the submarine industry [65]. Apart from enhancing classical transportation devices, inspirational novel ones were developed flowing the water strider's model. Prototypes of miniature robots which walk-in straight-line and function as water-pollutant monitors, displaying high transport capacities [66, 67]. Water collecting/storing systems are still developing as a solution for dry areas, starting from the Namib beetle's back special architecture [68].

Interdisciplinary researches on surface extreme wettability will be continued by discussing an intrinsically superhydrophobic behavior, characteristic for versatile structures entitled liquid marbles.

4. Small exponent—big impact: liquid marbles

4.1 State of the art

Liquid marbles are non-wettable structures, formed as a result of physical interactions between solid particles and a liquid drop. The formations are in fact represented by a liquid core covered in a particle shell (**Figure 6**) and exhibit a superhydrophobic-like behavior, without the intervention of surface modifications.

Among the first intents to obtain liquid marbles were carried out by Aussillous and Quéré [69], by rolling water droplets (1–10 mm^3) in a hydrophobic silica-covered *Lycopodium* powder bed (20 μm), as presented in **Figure 7**.

When compared to plain water drops, the manufactured liquid marbles did not wet the support, due to the fact that the liquid-solid interface (water-glass) is replaced with a solid-solid interface (*Lycopodium* particles-glass). They resemble raindrops falling on lotus leaves and collecting dust particles while rolling off, as previously discussed (The Lotus Effect) [70].

Liquid marbles' formulations are versatile, including various powders which differ in color, wetting degree, electrical charge, and even therapeutic activity.

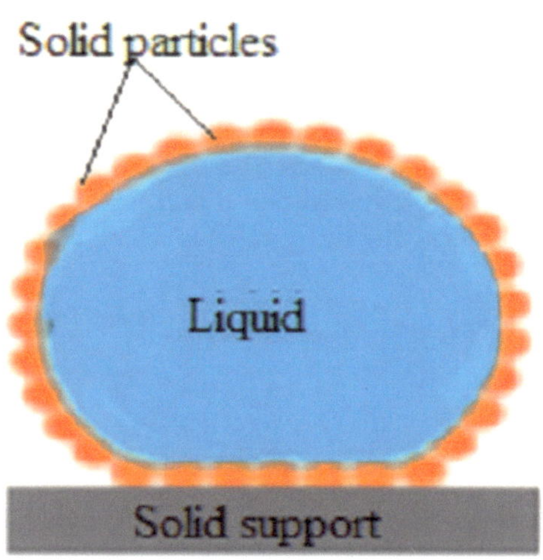

Figure 6.
Liquid marble structure.

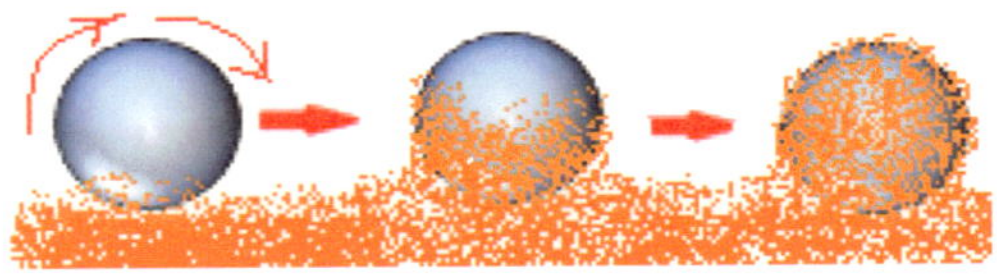

Figure 7.
Obtaining liquid marbles by rolling water drops into a Lycopodium powder bed.

Literature data indicates natural and synthetical powders such as *Lycopodium*, soot [71], respectively, polyvinylidene fluoride (PVDF), polytetrafluoroethylene (PTFE), polyethylene (PE) [72], polymethyl methacrylate (PMMA) [73], and hydrophobic copper powder [74]. The hydrophobic particle wall thickness varies depending on the particles, which are linked by van der Waals forces and distribute as mono- or a multilayers. In time, the wall undergoes changes depending on environmental conditions: multilayered shells correspond to "long surviving marbles," due to the possibility to stretch of particles surrounding the circumference, maintaining the system's integrity. Moreover, since large particles do not provide liquid core's flexibility and protection, nanoparticles are recommended as shell formers which fill in the gaps formed through compression [75].

Similar to superhydrophobic surfaces, liquid marbles can also exhibit special structural architectures, depending on their components. Particular cases include porous shells made of hydrophobic poly-high internal phase emulsion (HIPE) polymer, with particles interconnected by "gigapores" of micronic dimensions, resembling natural organisms like radiolarians (protozoa-producing mineral microtubes) or diatoms (microalgae with cells interconnected by tubes) [76]. After the $CuSO_4$ solution (core) evaporates through the shell, a $CuSO_4$ shell remains. The method is proposed as the model in designing spherical objects. Among liquid marbles with curious properties are the ones guided using electric fields which resemble Janus particles. They are obtained by forcing together two marbles with different shells, resulting in a bigger marble: half covered in carbon black and the other in Teflon (**Figure 8**) [77].

Liquid marble's interior phase usually includes high-surface tension liquids like water or glycerol, but literature data also suggests low tension liquids such as ethanol, methanol, toluene, hexadecane, and 1,4-dioxane [78]. It is possible for the shells' particles to remain at the liquid-gas interface or to be engulfed by the liquid core, resulting in stable marbles [72]. Other particular liquid marbles include Galinstan (eutectic liquid mixture of gallium, indium, tin) covered in Teflon, isolators (SiO_2), or semiconductors (CuO, ZnO, WO_3). They are resistant to high temperatures, float on water, but must be obtained in a diluted hydrochloric acid solution, as an unwanted reduction reaction takes place in air [79].

Cases of hydrophilic particle-covered liquid marbles are possible due to air trapped between particles, resulting in aggregates which cover the droplets [80].

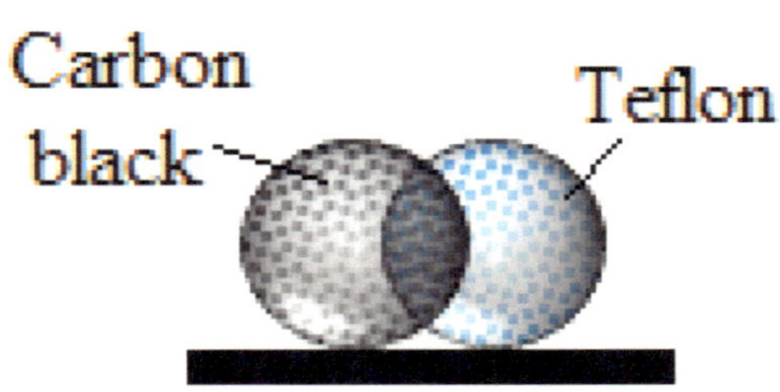

Figure 8.
Liquid marbles resembling Janus particles.

When discussing liquid marbles obtaining procedures, the most popular manufacturing method is the droplet rolling in a powder bed, as previously presented. Continuous research is developed concerning this domain since the proposed method is inefficient and time-consuming; irregularly covered marbles are formed and cannot be transposed at an industrial level. Methods including condensation and drop nucleation were recently reported: the liquid core is placed in a container, warmed by a heat source underneath. Hydrophobic particles (Cab-O-Sil fumed silica and micronic-sized Teflon) are distributed in a thin layer at the liquid-air interface. As the liquid boils, vapors condense and are covered by the particles. Micronic liquid marbles are formed. By heating these "parent-marbles," much smaller liquid droplets called "child-liquid marbles" are formed ("liquid marbles sweating"). The "child-marbles" roll off the "parent-marbles" and are more robust. Advantages of the method include industrially applicability of the technique and possibility to adapt conditions depending on the desired result [81]. Wrapping drops in transparent glass fibers, avoiding fluid evaporation, is a proposed design in developing new controlled drug release systems, water purification membranes [82]. Another automatized method is considered revolutionary by using instead of hydrophobic powders a superhydrophobic cloth of nanofibers. The drop is covered after impacting the cloth, resulting in highly resistant liquid marbles, with no internal phase loss [83].

Regarding their formulation, liquid marbles are versatile structures. The challenge is represented by choosing the appropriate components and experimental parameters of the fabrication/manufacturing process.

4.2 Liquid marbles: superhydrophobic entities with unique properties

Experimentally formed liquid marbles exhibit slightly different properties compared to naturally formed ones. Raindrops fall from big heights and get covered with particles due to internal currents and to kinetic energy [84]. Thus, the marbles exhibit *elastic-solid* and also *fluid* properties, known as a double "solid-fluid" character. The assumption that liquid marbles' elasticity is related to replacing liquid-solid (support) interface with solid (shell)-solid (support) interface is sustained by the absence of colored traces left by sodium hydroxide liquid marbles rolled on a phenolphthalein surface [85]. Moreover, shape changes occur for viscous marbles placed on an inclined plane: centrifugal forces determine marbles to slide off, while acceleration leads to the transformation of the spherical shape into "peanut," toroidal/"doughnut" shape [86], as presented in **Figure 9** [87].

Other experiments on liquid marbles' shape and elasticity proved how gradual compression of the marbles resulted in successive cracking and ultimately breaking of the shell, followed by collapse. Before the collapse, marbles allowed a compression up to 30% from the initial dimension [88].

Coalescence of the drops and possibility to engulf exterior objects may also be related to elasticity. As a result of applying exterior forces, two different liquid marbles connected through a glass bar undergo coalescence, forming a bigger structure and sharing a divided shell, as illustrated in **Figure 10(a)**. Regarding the possibility to "swallow" other objects, organic liquid covered in FD-POSS marbles is injected with another organic fluid. As long as the condition of immiscibility between the liquids is respected (proposed liquid pairs: toluene/DMSO, hexadecane/water), "encapsulating liquid marbles" are formed (**Figure 10 (b)**) [89].

Other liquid marbles' curious properties reside from freezing and drying in extreme temperature conditions. Experiments on PTFE-covered liquid marbles reveal surface aggregates, and multilayers are formed at the liquid-air interface, triggering wall thickening and shrinkage during *evaporation*. Thus, slow evaporation

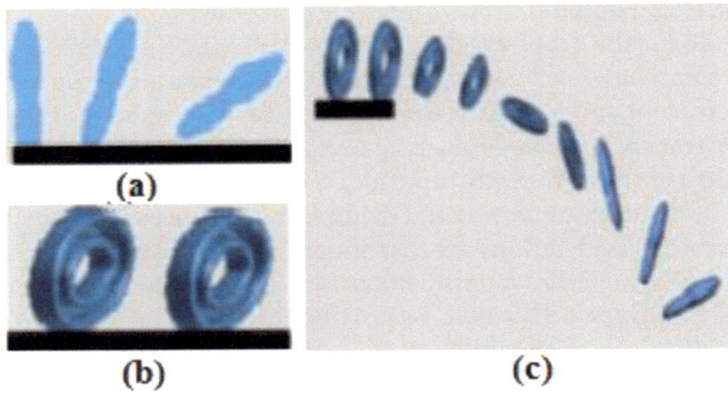

Figure 9.
Rolling liquid marbles: (a) "peanut" shape, (b) "doughnut" shape, and (c) transformation of "doughnut" into "peanut" shape, as the plane is removed [87].

Figure 10.
(a) Liquid marble coalescence: (b) FD-POSS liquid marble encapsulating DMSO.

of water results in prolonged resistance of the microparticle-covered marbles, with emerging applications in microfluidics [90]. The liquid marbles' shell layering raised curiosities: a mono-stratified shell determines the marbles to dry faster than a plain drop. The explanation lies in the fact that heat generates shrinkage at the liquid-air interface in case of the uncovered drop, while solid particles block interface compression during drying. Marbles covered in a multilayered shell dry harder than uncovered drops, depending on the thickness of the particle layer. Studies reveal the importance of environmental temperature and humidity in investigating evaporation: humid air delays evaporation of the liquid marble's internal phase [91, 92].

On the opposite pole of heating marbles are *freezing* ones, which were firstly reported as *Lycopodium*-covered water on a silicone support at −8°C. The marbles changed their shapes, as they flatten and extend sides: the "dome" shape evolves into a "flying saucer" shape, while the freezing process begins at the bottom and advances toward the top (**Figure 11**) [93].

Liquid marbles' behavior while *floating* on a liquid surface was also considered in recent experiments, since not only the liquid support surface deforms but also the marble itself. Particles covering the marble are packed between two fluids (liquid marble's core and carrier liquid) and change distribution leading to the marbles' collapse and release of the core into the support liquid. This phenomenon happens in normal conditions. In humid atmosphere, marbles maintain their shape many days while floating [94]. As expected, the deformation of the interface increases, consecutive to larger drops [95].

After floating investigations, *self-propelling* of liquid marbles became of interest, when an autonomous movement similar to Leidenfrost droplets was reported for water and alcohol marbles covered in extremely hydrophobic fumed silica. Supports include Petri dishes with water, and straight-line movement was observed. Taking into account the particles separating the core from the exterior, in the floating case,

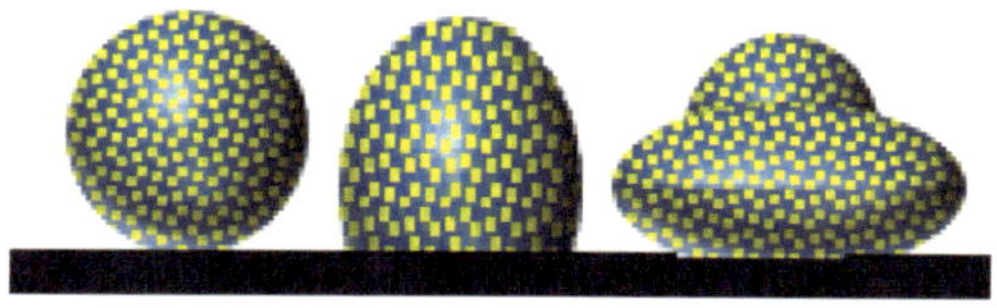

Figure 11.
Shape changes in freezing liquid marbles: Initial, "dome," and "flying saucer."

a vapor layer is responsible for marbles' support, similar to the Leidenfrost effect. The layer forms as ethanol evaporates (from the core). The Marangoni flow is triggered by ethanol condensation on the water support surface. Thus, the marbles begin to move without rolling. Fumed silica and Teflon-covered Janus marbles present no black traces while moving [96].

4.3 Small-scaled superhydrophobicity with innovative applications

Due to their versatile formulations and their special superhydrophobic-like properties, liquid marbles exhibit promising applications in various domains.

In the pharmaceutical domain, liquid marbles are known as precursors of hollow granules, microcapsules, and Pickering-like emulsions. Polytetrafluoroethylene (PTFE), aerosil, and Ballotini spheres as shells and binders (PVP, HPMC, HPC) are used to form liquid marbles which are dried through various methods: moist air at 24°C, freezing at −50°C, and dry air at 60°C, 80°C, 100°C). *Hollow granule* formation is promoted by high drying temperatures, nanometric particles, and high binder concentrations. Therefore, aerosil proved successful regarding spherical shape generation, when used together with HPMC and drying at 100°C forming ideal-shaped hollow granules [97], as presented in **Figure 12** [87].

Moreover, liquid marbles are able to include low solubility and hydrophobic active ingredients, representing formulation alternatives in case of substance incompatibilities and targeted release drugs (e.g., intestine and not stomach). The active ingredient's protection is mandatory against local acidity/enzymes and pathogens/other substances competing for binding sites and can be achieved by choosing the ideal development process while following Quality by Design Guidelines [87].

Microcapsules can also be obtained from liquid marbles: exposed to solvent vapors, submicrometer-sized polystyrene particles (PDEA-PS) covering

Figure 12.
Hollow granule [87].

poly(2(diethylamino)ethyl methacrylate) cores undergo a polymerization reaction, forming a filled capsule. After the core evaporates, the now empty microcapsule maintains its shape, representing an important idea for further design of modified drug release systems [98].

Liquid marbles have also been reported as precursors of *Pickering-like emulsions*. The difference between classical and Pickering emulsions stands in the lack of added stabilizing agents. Pickering emulsions are stabilized by solid particles adsorbed at the internal and external phase interface. These adsorbed particles are attached to the drop they cover and are wetted in both the watery and the oily phases, conferring integrity to the immersed drop [99, 100]. Stabilizer particles include clays, latexes, calcium carbonate, carbon black, magnetic particles, proteins, and even bacteria. Hydrophobic particles stabilize water/oil emulsions, and hydrophilic ones stabilize oi/water emulsions [99]. Stable Pickering like emulsions containing *Lycopodium*-covered liquid marbles were obtained in PDMS, as presented in **Figure 13**.

Experiments show good stability of marbles immersed in less polar liquids (silicone fluids, aromatic solvents), while collapse is a trigger in polar solvents. Pickering-like emulsions find their applicability in cosmetic formulations due to no allergenic, cytotoxic, or hemolytic stabilizers. Topical use of caffeine Pickering emulsions in controlled studies revealed higher absorption than the other pharmaceutical forms, due to silica-covered liquid marbles, which promote epidermal caffeine absorption from the aqueous phase of the emulsion [101]. Also, retinol included in the oily phase of a Pickering-like emulsion is stabilized against UV radiation and only penetrates the corneous layers of the epidermis [100].

Sticking to the field of topical application, liquid marble formulations represent a basis in foundation, antiperspirants/deodorants, solar protection products, and some drug formulation. Easy application is followed by a moisture and cooling sensation due to internal phase liberation. Among components, the most popular are deionized/floral water (50–90%) mixed with polymers/copolymers (PVP), wetting agents (hyaluronic acid), hydrosoluble vitamins, preservatives, and antioxidants. Such formulations are recommended for oily skins, due to a low oil content (<10%). Therapeutic agents may be added: antibacterial, antifungal, analgesic, keratolytic, corticosteroids, etc.

A novelty in blood typing is represented by liquid marbles as *biological microreactors*: hydrophobic-precipitated $CaCO_3$-covered blood drops are injected with antibody solutions (anti-A, anti-B, anti-D). If the color changes from red to dark

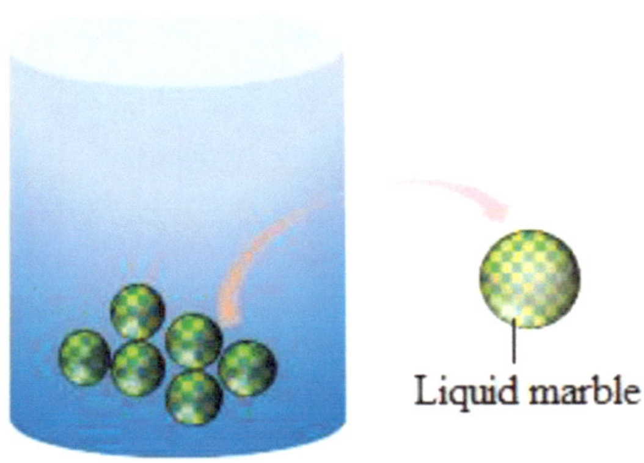

Figure 13.
Pickering-like emulsion.

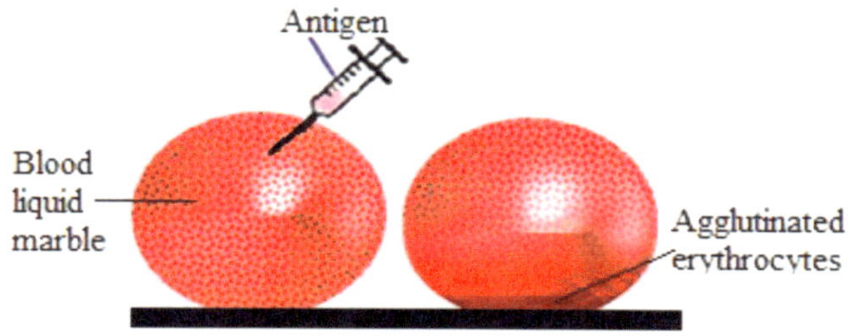

Figure 14.
Hemagglutination reaction inside a blood liquid marble [87].

red, the hemagglutination reaction is considered positive, as illustrated in **Figure 14**, and the blood type is immediately identified. If the color does not change, the reaction is negative, and another drop is tested using another antibody. The technique is considered innovative: low contamination risks and reagent costs and necessity of small blood samples, which is beneficial for patients [102].

PTFE liquid marbles were also used as *cell culturing medium*: spheroids successfully developed from $HepG_2$ hepatocellular cancer cells. Advantages include promoting cell aggregation due to restricted space and no human intervention. In order to screen cell development inside the marble, magnetic particles are proposed for shells, leaving an observation section open when a magnetic field is near. These methods are used in cell physiology screening or tissue engineering [103], stem cells evolution into embryoid bodies [104], and bacterial culturing especially for anaerobic species [105].

Liquid marbles are providers of 3D spherical space with adjustable volume and formulation and can also be regarded as *chemical micro-reactors*, hosting different chemical reactions resulting in toxic/explosive unwanted products, in small amounts, and isolated. Intervention from the outside is possible for "intelligent marbles" covered in magnetic particles, in order to inoculate a new reagent into the reaction, collect a product, identify, or quantitatively evaluate a certain compound. Indicators of core chemical reactions include color changes, chemiluminescence, and precipitation reactions [106]. Besides hosting chemical reactions, some marbles called "plasmonic liquid marbles" are covered in Ag/Au nanoparticles and represent special analytical platform precursors. They function as qualitative and also quantitative detectors for waste products resulted from industrial spills, being able to trace compounds in concentration of femto- or ato-molar concentrations (10^{-15}, 10^{-18}) [107]. "Cleaning agent stimuli-responsive liquid marbles" detect pollutants and signal their presence by shell breakage. The core eliminates a detoxifying agent (1 N-oxone covered in Cab-O-Sil T-530 shells) which cleans oil-contaminated water [108].

5. Conclusions

This chapter is an interdisciplinary approach on extreme wettability, granting particular attention to superhydrophobic natural and artificial surfaces and to liquid marbles, as exponent. Literature data is reunited in order to offer a unique and complex understanding of superficial properties from a theoretical point of view, in correlation with examples from the natural environment. An extensive picture illustrates how superhydrophobicity was initially interpreted, how its understanding evolved, becoming of large exploitation in many industrial fields. Superficial properties and liquid marbles are linked through conceptual similarities, as an opening gate to numerous applications.

Liquid marble exploration substantially advanced during the last years, from the phase of basic understanding through wetting models to more complex interpretations, obtaining methods and applications. Studies revealed a "non-wetting" contact with solid supports and many unexpected properties, such as versatility in choice of cores and shells, recoverable deformability, ability to float on water, low evaporation rate, and significant advantages derived from a well-confined compartment. Emerging applications discussed in this chapter are diverse and offer a rich variety of further exploitation possibilities, arising from complex structural designs.

Conflict of interest

We, the authors of this paper Cristina Elena Dinu-Pîrvu, Roxana-Elena Avrămescu, Mihaela Violeta Ghica, and Lăcrămioara Popa, declare no conflicts of interests.

Author details

Cristina Elena Dinu-Pîrvu, Roxana-Elena Avrămescu, Mihaela Violeta Ghica* and Lăcrămioara Popa
Faculty of Pharmacy, University of Medicine and Pharmacy "Carol Davila", Bucharest, Romania

*Address all correspondence to: mihaela.ghica@umfcd.ro

References

[1] Vincent JF, Bogatyreva OA, Bogatyrev NR, Bowyer A, Pahl AK. Biomimetics: Its practice and theory. Journal of the Royal Society, Interface. 2006;**3**(9):471-482. DOI: 10.1098/rsif.2006.0127

[2] Vincent J. Biomimetics–A review. Proceedings of the Institution of Mechanical Engineers. Part H, Journal of Engineering in Medicine. 2009;**223**:919-939

[3] Junlong S, Rojas OJ. Approaching superhydrophobicity from cellulosic materials. A review. Nordic Pulp & Paper Research Journal. 2013;**28**(2):216-238. DOI: 10.3183/NPPRJ-2013-28-02-p216-238

[4] Wenzel RN. Resistance of solid surfaces to wetting by water. Industrial and Engineering Chemistry. 1936;**28**(8):988-994. DOI: 10.1021/ie50320a024

[5] Cassie ABD, Baxter S. Wettability of porous surfaces. Transactions of the Faraday Society. 1944;**40**:546. DOI: 10.1039/tf9444000546

[6] Lafuma A, Quéré D. Superhydrophobic states. Nature Materials. 2003;**2**(7):457. DOI: 10.1038/nmat924

[7] Quéré D. Model droplets. Nature Materials. 2004;**3**(2):79. DOI: 10.1038/nmat1062

[8] Bormashenko E. Progress in understanding wetting transitions on rough surfaces. Advances in Colloid and Interface Science. 2015;**222**:92-103. DOI: 10.1016/j.cis.2014.02.009

[9] Li Y, Quéré D, Lv C, Zheng Q. Monostable superrepellent materials. Proceedings of the National Academy of Sciences of the United States of America. 2017;**114**(13):3387-3392. DOI: 10.1073/pnas.1614667114

[10] Tian Y, Su B, Jiang L. Interfacial material system exhibiting Superwettability. Advanced Materials. 2014;**26**(40):6872-6897. DOI: 10.1002/adma.201400883

[11] Guo Z, Liu W, Su BL. Superhydrophobic surfaces: From natural to biomimetic to functional. Journal of Colloid and Interface Science. 2011;**353**(2):335-355. DOI: 10.1016/j.jcis.2010.08.047

[12] Gu Z-Z, Wei H-M, Zhang R-Q, Han G-Z, Pan C, Zhang H, et al. Artificial silver ragwort surface. Applied Physics Letters. 2005;**86**(20):201915. DOI: 10.1063/1.1931054

[13] Brewer SA, Willis CR. Structure and oil repellency. Textiles with liquid repellency to hexane. Applied Surface Science. 2008;**254**(20):6450-6454. DOI: 10.1016/j.apsusc.2008.04.053

[14] Brinker CJ, Branson E, Kissel DJ, Cook A, Singh S. Superhydrophobic coating. USPatent 2008, (US7485343 B1), US7485343 B1-US7485343 B1

[15] Yong J, Chen F, Yang Q, Huo J, Hou X. Superoleophobic surfaces. Chemical Society Reviews. 2017;**46**(14):4168-4217. DOI: 10.1039/C6CS00751A

[16] Chen H, Zhang P, Zhang L, Liu H, Jiang Y, Zhang D, et al. Continuous directional water transport on the peristome surface of Nepenthes alata. Nature. 2016;**532**(7597):85-89. DOI: 10.1038/nature17189

[17] Ueda E, Levkin PA. Emerging applications of superhydrophilic-superhydrophobic micropatterns. Advanced Materials. 2013;**25**(9):1234-1247. DOI: 10.1002/adma.201204120

[18] Mayser M, Bohn H, Reker M, Barthlott W. Measuring air layer volumes retained by submerged floating-ferns Salvinia and biomimetic superhydrophobic surfaces. Beilstein Journal of Nanotechnology. 2014;**5**:812-821

[19] Verboven P, Pedersen O, Ho QT, Nicolai BM, Colmer TD. The mechanism of improved aeration due to gas films on leaves of submerged rice. Plant, Cell & Environment. 2014;**37**(10):2433-2452. DOI: 10.1111/pce.12300

[20] Winkel A, Visser EJW, Colmer TD, Brodersen KP, Voesenek LACJ, Sand-Jensen K, et al. Leaf gas films, underwater photosynthesis and plant species distributions in a flood gradient. Plant, Cell & Environment. 2016;**39**(7):1537-1548. DOI: 10.1111/pce.12717

[21] Feng L, Zhang Y, Xi J, Zhu Y, Wang N, Xia F, et al. Petal effect: A superhydrophobic state with high adhesive force. Langmuir. 2008;**24**(8):4114-4119. DOI: 10.1021/la703821h

[22] Avramescu RE, Ghica MV, Dinu-Pirvu C, Prisada R, Popa L. Superhydrophobic natural and artificial surfaces-a structural approach. Materials (Basel, Switzerland). 2018;**11**(5):1-24. DOI: 10.3390/ma11050866

[23] Savage N. Synthetic coatings: Super surfaces. Nature. 2015;**519**(7544):S7-S9. DOI: 10.1038/519S7a

[24] Liu K, Du J, Wu J, Jiang L. Superhydrophobic gecko feet with high adhesive forces towards water and their bio-inspired materials. Nanoscale. 2012;**4**(3):768-772. DOI: 10.1039/C1NR11369K

[25] Scott AR. Polymers: Secrets from the deep sea. Nature. 2015;**519**(7544):S12-S13. DOI: 10.1038/519S12a

[26] Wagner T, Neinhuis C, Barthlott W. Wettability and contaminability of insect wings as a function of their surface sculptures. Acta Zoologica. 1996;**77**(3):213-225. DOI: 10.1111/j.1463-6395.1996.tb01265.x

[27] Choi W, Tuteja A, Mabry JM, Cohen RE, McKinley GH. A modified Cassie-Baxter relationship to explain contact angle hysteresis and anisotropy on non-wetting textured surfaces. Journal of Colloid and Interface Science. 2009;**339**(1):208-216. DOI: 10.1016/j.jcis.2009.07.027

[28] Kazufumi O, Mamoru S, Yusuke T, Ichiro N. Development of a transparent and ultrahydrophobic glass plate. Japanese Journal of Applied Physics. 1993;**32**(4B):L614

[29] Shibuichi S, Onda T, Satoh N, Tsujii K. Super water-repellent surfaces resulting from fractal structure. The Journal of Physical Chemistry. 1996;**100**(50):19512-19517. DOI: 10.1021/jp9616728

[30] Onda T, Shibuichi S, Satoh N, Tsujii K. Super-water-repellent fractal surfaces. Langmuir. 1996;**12**(9):2125-2127. DOI: 10.1021/la950418o

[31] Chen W, Fadeev AY, Hsieh MC, Öner D, Youngblood J, McCarthy TJ. Ultrahydrophobic and ultralyophobic surfaces: Some comments and examples. Langmuir. 1999;**15**(10):3395-3399. DOI: 10.1021/la990074s

[32] Öner D, McCarthy TJ. Ultrahydrophobic surfaces. Effects of topography length scales on wettability. Langmuir. 2000;**16**(20):7777-7782. DOI: 10.1021/la000598o

[33] Cheng Y-T, Rodak DE, Angelopoulos A, Gacek T. Microscopic observations of condensation of water on lotus leaves. Applied Physics Letters. 2005;**87**(19):194112. DOI: 10.1063/1.2130392

[34] Li N, Xia T, Heng L, Liu L. Superhydrophobic Zr-based metallic glass surface with high adhesive force. Applied Physics Letters. 2013;**102**(25):251603. DOI: 10.1063/1.4812480

[35] Li S, Xie H, Zhang S, Wang X. Facile transformation of hydrophilic cellulose into superhydrophobic cellulose. Chemical Communications. 2007;**46**:4857. DOI: 10.1039/b712056g

[36] Barthlott W, Neinhuis C. Purity of the sacred lotus, or escape from contamination in biological surfaces. Planta. 1997;**202**:1-8

[37] Hoefnagels HF, Wu D, De With G, Ming W. Biomimetic superhydrophobic and highly oleophobic cotton textiles. Langmuir. 2007;**23**(26):13158-13163. DOI: 10.1021/la702174x

[38] Bartell FE, Shepard JW. Surface roughness as related to hysteresis of contact angles. II. The systems paraffin-3 molar calcium chloride solution-air and paraffin-glycerol-air. The Journal of Physical Chemistry. 1953;**57**(4):455-458. DOI: 10.1021/j150505a015

[39] Norton FJ. U.S. Patent. 1945, Oct 9

[40] Frederick KJ. Performance and problems of pharmaceutical suspensions. Journal of Pharmaceutical Sciences. 1961;**50**(6):531-535. DOI: 10.1002/jps.2600500619

[41] Pike N, Richard D, Foster W, Mahadevan L. How aphids lose their marbles. Proceedings. Biological Sciences/The Royal Society. 2002;**269**(1497):1211-1215. DOI: 10.1098/rspb.2002.1999

[42] Ravanetti, F, Caccioli A. Primary Osteintegration in the Study of Biomimetic Surfaces.pdf. On Biomimetics. Croatia: InTech pp. 91-106. ISBN: 978-953-307-271-5

[43] Pickering SU. Pickering: Emulsions. Journal of the Chemical Society. 1907;**91**:2001-2021. DOI: 10.1039/ct9079102001

[44] Wang J, Wen Y, Hu J, Song Y, Jiang L. Fine control of the wettability transition temperature of colloidal-crystal films: From superhydrophilic to superhydrophobic. Advanced Functional Materials. 2007;**17**(2):219-225. DOI: 10.1002/adfm.200600101

[45] Mozumder MS, Zhang H, Zhu J. Mimicking lotus leaf: Development of micro-nanostructured biomimetic Superhydrophobic polymeric surfaces by ultrafine powder coating technology. Macromolecular Materials and Engineering. 2011;**296**(10):929-936. DOI: 10.1002/mame.201100080

[46] Genzer J, Efimenko K. Creating long-lived Superhydrophobic polymer surfaces through mechanically assembled monolayers. Science. 2000;**290**(5499):2130-2133. DOI: 10.1126/science.290.5499.2130

[47] Song. Approaching super-hydrophobicity from cellulosic materials: A review. Nordic Pulp & Paper Research Journal. 2013;**28**(2):216-238. DOI: 10.3183/NPPRJ-2013-28-02-p216-238

[48] Wang Z, Shen X, Qian T, Wang J, Sun Q, Jin C. Facile fabrication of a PDMS@stearic acid-kaolin coating on lignocellulose composites with Superhydrophobicity and flame Retardancy. Materials. 2018;**11**(5):1-10. DOI: 10.3390/ma11050727

[49] Chatzigrigoriou A, Manoudis Panagiotis N, Karapanagiotis I. Fabrication of water repellent coatings using waterborne resins for the protection of the cultural heritage. Macromolecular Symposia. 2013;**331-332**(1):158-165. DOI: 10.1002/masy.201300063

[50] Wong TS, Kang SH, Tang SK, Smythe EJ, Hatton BD, Grinthal A, et al. Bioinspired self-repairing slippery surfaces with pressure-stable omniphobicity. Nature. 2011;**477**(7365):443-447. DOI: 10.1038/nature10447

[51] Dolgin E. Textiles: Fabrics of life. Nature. 2015;**519**(7544):S10-S11. DOI: 10.1038/519S10a

[52] Yao X, Song Y, Jiang L. Applications of bio-inspired special wettable surfaces. Advanced Materials. 2011;**23**(6):719-734. DOI: 10.1002/adma.201002689

[53] Fan H, Chen P, Qi R, Zhai J, Wang J, Chen L, et al. Greatly improved blood compatibility by microscopic multiscale design of surface architectures. Small. 2009;**5**(19):2144-2148. DOI: 10.1002/smll.200900345

[54] Tomšič B, Simončič B, Orel B, Černe L, Tavčer PF, Zorko M, et al. Sol–gel coating of cellulose fibres with antimicrobial and repellent properties. Journal of Sol-Gel Science and Technology. 2008;**47**(1):44-57. DOI: 10.1007/s10971-008-1732-1

[55] Epstein AK, Wong TS, Belisle RA, Boggs EM, Aizenberg J. Liquid-infused structured surfaces with exceptional anti-biofouling performance. Proceedings of the National Academy of Sciences. 2012;**109**(33):13182-13187. DOI: 10.1073/pnas.1201973109

[56] Geyer FL, Ueda E, Liebel U, Grau N, Levkin PA. Superhydrophobic-superhydrophilic micropatterning: Towards genome-on-a-chip cell microarrays. Angewandte Chemie. 2011;**50**(36):8424-8427. DOI: 10.1002/anie.201102545

[57] Lai Y, Chen Z, Lin C. Recent Progress on the Superhydrophobic surfaces with special adhesion: From natural to biomimetic to functional. Journal of Nanoengineering and Nanomanufacturing. 2011;**1**:18-34

[58] Lee YY, Narayanan K, Gao SJ, Ying JY. Elucidating drug resistance properties in scarce cancer stem cells using droplet microarray. Nano Today. 2012;**7**(1):29-34. DOI: 10.1016/j.nantod.2012.01.003

[59] Kim P, Wong TS, Alvarenga J, Kreder MJ, Adorno-Martinez WE, Aizenberg J. Liquid-infused nanostructured surfaces with extreme anti-ice and anti-frost performance. ACS Nano. 2012;**6**(8):6569-6577. DOI: 10.1021/nn302310q

[60] Nakajima A, Abe K, Hashimoto K, Watanabe T. Preparation of hard super-hydrophobic films with visible light transmission. Thin Solid Films. 2000;**376**(1-2):140-143. DOI: 10.1016/s0040-6090(00)01417-6

[61] Meuler AJ, Smith JD, Varanasi KK, Mabry JM, McKinley GH, Cohen RE. Relationships between water wettability and ice adhesion. ACS Applied Materials & Interfaces. 2010;**2**(11):3100-3110. DOI: 10.1021/am1006035

[62] Yabu H, Shimomura M. Single-step fabrication of transparent Superhydrophobic porous polymer films. Chemistry of Materials. 2005;**17**(21):5231-5234. DOI: 10.1021/cm051281i

[63] Vasiljević J, Gorjanc M, Tomšič B, Orel B, Jerman I, Mozetič M, et al. The surface modification of cellulose fibres to create super-hydrophobic, oleophobic and self-cleaning properties. Cellulose. 2013;**20**(1):277-289. DOI: 10.1007/s10570-012-9812-3

[64] Wen L, Tian Y, Jiang L. Bioinspired super-wettability from fundamental research to practical applications. Angewandte Chemie. 2015;**54**(11):3387-3399. DOI: 10.1002/anie.201409911

[65] Bhushan B. Bioinspired structured surfaces. Langmuir. 2012;**28**(3): 1698-1714. DOI: 10.1021/la2043729

[66] Koh JS, Yang E, Jung GP, Jung SP, Son JH, Lee SI, et al. Jumping on water: Surface tension-dominated jumping of water striders and robotic insects. Science. 2015;**349**(6247):517-521. DOI: 10.1126/science.aab1637

[67] Zhao J, Zhang X, Chen N, Pan Q. Why superhydrophobicity is crucial for a water-jumping microrobot? Experimental and theoretical investigations. ACS Applied Materials and Interfaces. 2012;**4**(7):3706-3711. DOI: 10.1021/am300794z

[68] Zhai L, Berg MC, Cebeci FÇ, Kim Y, Milwid JM, Rubner MF, et al. Patterned superhydrophobic surfaces: Toward a synthetic mimic of the namib desert beetle. Nano Letters. 2006;**6**(6):1213-1217. DOI: 10.1021/nl060644q

[69] Aussillous P, Quéré D. Liquid marbles. Nature. 2001;**411**(6840):924-927. DOI: 10.1038/35082026

[70] Mahadevan L. Non-stick water. Nature. 2001;**411**(June):896-895. DOI: 10.1038/35082164

[71] Quéré D, Aussillous P. Properties of liquid marbles. Proceedings of the Royal Society A: Mathematical, Physical and Engineering Sciences. 2006;**2067**(462):973-999. DOI: 10.1098/rspa.2005.1581

[72] Bormashenko E, Balter R, Aurbach D. Formation of liquid marbles and wetting transitions. Journal of Colloid and Interface Science. 2012;**384**(1):157-161. DOI: 10.1016/j.jcis.2012.06.023

[73] McEleney P, Walker GM, Larmour IA, Bell SEJ. Liquid marble formation using hydrophobic powders. Chemical Engineering Journal. 2009;**147**(2-3): 373-382. DOI: 10.1016/j.cej.2008.11.026

[74] Walker GM, McEleney P, Al-Muhtaseb AAH, Bell SEJ. Liquid marble granulation using super-hydrophobic powders. Chemical Engineering Journal. 2013;**228**:984-992. DOI: 10.1016/j.cej.2013.05.055

[75] Nguyen T, Shen W, Hapgood K. Observation of the liquid marble morphology using confocal. Microscopy. 2010;**162**:396-405

[76] Gokmen MT, Dereli B, De Geest BG, Du Prez FE. Complexity from simplicity: Unique polymer capsules, rods, monoliths, and liquid marbles prepared via HIPE in microfluidics. Particle & Particle Systems Characterization. 2013;**30**(5):438-444. DOI: 10.1002/ppsc.201200154

[77] Bormashenko E, Bormashenko Y, Pogreb R, Gendelman O. Janus droplets: Liquid marbles coated with dielectric/semiconductor particles. Langmuir. 2011;**27**:7-10

[78] Matsukuma D, Watanabe H, Yamaguchi H, Takahara A. Preparation of low-surface-energy poly[2-(perfluorooctyl)ethyl acrylate] microparticles and its application to liquid marble formation. Langmuir. 2011;**27**(4):1269-1274. DOI: 10.1021/la1040689

[79] Sivan V, Tang S-Y, O'Mullane AP, Petersen P, Eshtiaghi N, Kalantar-zadeh K, et al. Liquid metal marbles. Advanced Functional Materials. 2013;**23**(2):144-152. DOI: 10.1002/adfm.201200837

[80] Janardan N, Panchagnula M, Bormashenko E. Liquid marbles: Physics and applications. Sadhana - Academy Proceedings in Engineering Sciences. 2015;**40**(3):653-671. DOI: 10.1007/s12046-015-0365-7

[81] Bhosale PS, Panchagnula MV. Sweating liquid micro-marbles: Dropwise condensation on hydrophobic

nanoparticulate materials. Langmuir. 2012;**28**(42):14860-14866. DOI: 10.1021/la303133y

[82] Zuber K, Evans D, Murphy P. Nanoporous glass films on liquids. ACS Applied Materials & Interfaces. 2014;**6**(1):507-512. DOI: 10.1021/am404570a

[83] Mele E, Bayer IS, Nanni G, Heredia-Guerrero JA, Ruffilli R, Ayadi F, et al. Biomimetic approach for liquid encapsulation with nanofibrillar cloaks. Langmuir. 2014;**30**(10):2896-2902. DOI: 10.1021/la4048177

[84] Whitby CP, Bian X, Sedev R. Spontaneous liquid marble formation on packed porous beds. Soft Matter. 2012;**8**:1-29

[85] Bormashenko E, Bormashenko Y, Musin A, Barkay Z. On the mechanism of floating and sliding of liquid marbles. Chemphyschem : A European Journal of Chemical Physics and Physical Chemistry. 2009;**10**(4):654-656. DOI: 10.1002/cphc.200800746

[86] Aussillous P, QuÉRÉ D. Shapes of rolling liquid drops. Journal of Fluid Mechanics. 2004;**512**:133-151. DOI: 10.1017/S0022112004009747

[87] Avrămescu R-E, Ghica M-V, Dinu-Pîrvu C, Udeanu D, Popa L. Liquid marbles: From industrial to medical applications. Molecules. 2018;**23**(5):1-30. DOI: 10.3390/molecules23051120

[88] Asare-Asher S, Connor JN, Sedev R. Elasticity of liquid marbles. Journal of Colloid and Interface Science. 2015;**449**:341-346. DOI: 10.1016/j.jcis.2015.01.067

[89] Bormashenko E, Balter R, Aurbach D, Anton S, Viktor V, Vladimir S. Liquid marbles swallowing one another and extraneous objects. 2014. pp. 1-14

[90] Tosun A, Erbil HY. Evaporation rate of PTFE liquid marbles. Applied Surface Science. 2009;**256**(5):1278-1283. DOI: 10.1016/j.apsusc.2009.10.035

[91] Erbil HY. Evaporation of pure liquid sessile and spherical suspended drops: A review. Advances in Colloid and Interface Science. 2012;**170**(1-2):67-86. DOI: 10.1016/j.cis.2011.12.006

[92] Laborie B, Lachaussée F, Lorenceau E, Rouyer F. How coatings with hydrophobic particles may change the drying of water droplets: Incompressible surface versus porous media effects. Soft Matter. 2013;**9**(19):4822-4830. DOI: 10.1039/c3sm50164g

[93] Hashmi A, Strauss A, Xu J. Freezing of a liquid marble. Langmuir. 2012;**28**(28):10324-10328. DOI: 10.1021/la301854f

[94] Ooi CH, Vadivelu RK, St John J, Dao DV, Nguyen N-T. Deformation of a floating liquid marble. Soft Matter. 2015;**11**(23):4576-4583. DOI: 10.1039/C4SM02882A

[95] Bormashenko E, Bormashenko Y, Musin A. Water rolling and floating upon water: Marbles supported by a water/marble interface. Journal of Colloid and Interface Science. 2009;**333**(1):419-421. DOI: 10.1016/j.jcis.2008.09.079

[96] Bormashenko E, Bormashenko Y, Grynyov R, Aharoni H, Whyman G, Binks BP. Self-propulsion of liquid marbles: Leidenfrost-like levitation driven by Marangoni flow. The Journal of Physical Chemistry C. 2015;**119**(18):9910-9915. DOI: 10.1021/acs.jpcc.5b01307

[97] Khanmohammadi B, Yeo L, Hapgood KP. Formation of hollow granules from hydrophobic powders. 2007(1989):451-459. Available from:

https://www.researchgate.net/publication/228533191_Formation_of_hollow_granules_from_hydrophobic_powders

[98] Ueno K, Hamasaki S, Wanless EJ, Nakamura Y, Fujii S. Microcapsules fabricated from liquid marbles stabilized with latex particles. Langmuir. 2014;**30**(11):3051-3059. DOI: 10.1021/la5003435

[99] Binks BP. Particles as surfactants—Similarities and differences. Current Opinion in Colloid & Interface Science. 2002;**7**(1-2):21-41. DOI: 10.1016/s1359-0294(02)00008-0

[100] Chevalier Y, Bolzinger M-A. Emulsions stabilized with solid nanoparticles: Pickering emulsions. Colloids and Surfaces A: Physicochemical and Engineering Aspects. 2013;**439**:23-34. DOI: 10.1016/j.colsurfa.2013.02.054

[101] Frelichowska J, Bolzinger MA, Valour JP, Mouaziz H, Pelletier J, Chevalier Y. Pickering w/o emulsions: Drug release and topical delivery. International Journal of Pharmaceutics. 2009;**368**(1-2):7-15. DOI: 10.1016/j.ijpharm.2008.09.057

[102] Arbatan T, Li L, Tian J, Shen W. Liquid marbles as micro-bioreactors for rapid blood typing. Advanced Healthcare Materials. 2012;**1**(1):80-83. DOI: 10.1002/adhm.201100016

[103] Arbatan T, Al-Abboodi A, Sarvi F, Chan PPY, Shen W. Tumor inside a pearl drop. Advanced Healthcare Materials. 2012;**1**(4):467-469. DOI: 10.1002/adhm.201200050

[104] Sarvi F, Arbatan T, Chan PPY, Shen W. A novel technique for the formation of embryoid bodies inside liquid marbles. RSC Advances. 2013;**3**(34):14501-14501. DOI: 10.1039/c3ra40364e

[105] Tian J, Fu N, Chen X. D.; Shen, W., respirable liquid marble for the cultivation of microorganisms. colloids and surfaces. B. Biointerfaces. 2013;**106**:187-190. DOI: 10.1016/j.colsurfb.2013.01.016

[106] Xue Y, Wang H, Zhao Y, Dai L, Feng L, Wang X, et al. Magnetic liquid marbles: A "precise" miniature reactor. Advanced Materials. 2010;**22**(43):1-5. DOI: 10.1002/adma.201001898

[107] Lee HK, Lee YH, Phang IY, Wei J, Miao YE, Liu T, et al. Plasmonic liquid marbles: A miniature substrate-less SERS platform for quantitative and multiplex ultratrace molecular detection. Angewandte Chemie (International Ed. in English). 2014;**53**(20):5054-5058. DOI: 10.1002/anie.201401026

[108] Hoffman DM, Chiu IL. Solid-water detoxifying reagents for chemical and biological reagents-United States Patent (45). 2006;**2**(12):5-8. Available from: https://patents.google.com/patent/US20030160209